THE BOOK OF KNIVES

YVAN A. DE RIAZ

THE BOOK OF KNIVES

CROWN PUBLISHERS, INC.
NEW YORK

Printed in Italy

First published in the United States of America by
Crown Publishers, Inc.

Inquiries should be addressed to
Crown Publishers, Inc.
One Park Avenue
New York, New York 10016

Published simultaneously in Canada by
General Publishing Company Limited

Library of Congress Cataloging in Publication Data
De Riaz, Yvan.
The book of knives.
1. Knives. I. Title.
TS380.D4 1982 621.9'32 81-22057
ISBN 0-517-54361-3 AACR2

10 9 8 7 6 5 4 3 2 1

CONTENTS

FOREWORD

Of all every day tools, the knife is one of the most ancient. It is also one whose history and manufacture are little-known except to specialists. Recently, however, a world-wide renewal of interest in knives has become apparent. This new popularity originated in the United States where, since 1945, a high level of craftsmanship in knifemaking has sprung up anew.

I have tried to present as complete a picture as possible of the world of cutlery, and the reader will find in this book the old as well as the new. Nonetheless, the 19th and 20th centuries have been given more attention than other periods, as being generally more interesting to enthusiasts and collectors today. Besides technical information, I have also tried to bring out other aspects of the subject: the history, style, and uses of various knifes, always seen within the general context.

Knives may vary according to their use, their size, the form of the blade and the handle, the kind of tang, and the materials of which they are made. If all of the possible combinations are taken into account, there could be tens, even hundreds of thousands of different kinds of knives. The purpose of this book is not to be a catalogue. Rather, the examples have been selected on the basis of their variety, their technical or historical interest, their appeal as novelties, or simply their beauty.

The preparation of this book has brought me into contact with 211 people and institutions — manufacturers, craftsmen, knife-sellers, collectors and museum keepers — in 14 countries. It is impossible to name them all here, but at the end of this book there is a word of thanks and acknowledgement for all those who helped make the book possible. However, I would like to express my special gratitude to five people who have been particularly helpful in the making of this book, either through the documents and illustrations they have placed at my disposal or through the advice they have given: Buster Warensky, President of the American Knifemakers Guild at the time of publication; Paul-Louis Frachebourg and Monique Fillettaz of the Coutellerie du Mont Blanc in Geneva; Gabriel Besson, cutler, Lausanne; Robert Barincou, collector, Paris.

I would also like to salute the American craftsmen and to thank them for their goodwill towards a stranger... who soon became a friend.

Yvan A. de Riaz

The prehistoric knife was an all-purpose tool. At first, it could hardly be distinguished from other stone tools, as can be seen from this illustration. At left is a knife, then a point at the end of a 'stalk', a notched point and a leaf-like point. Antiquités nationales, Saint-Germain-en-Laye, France

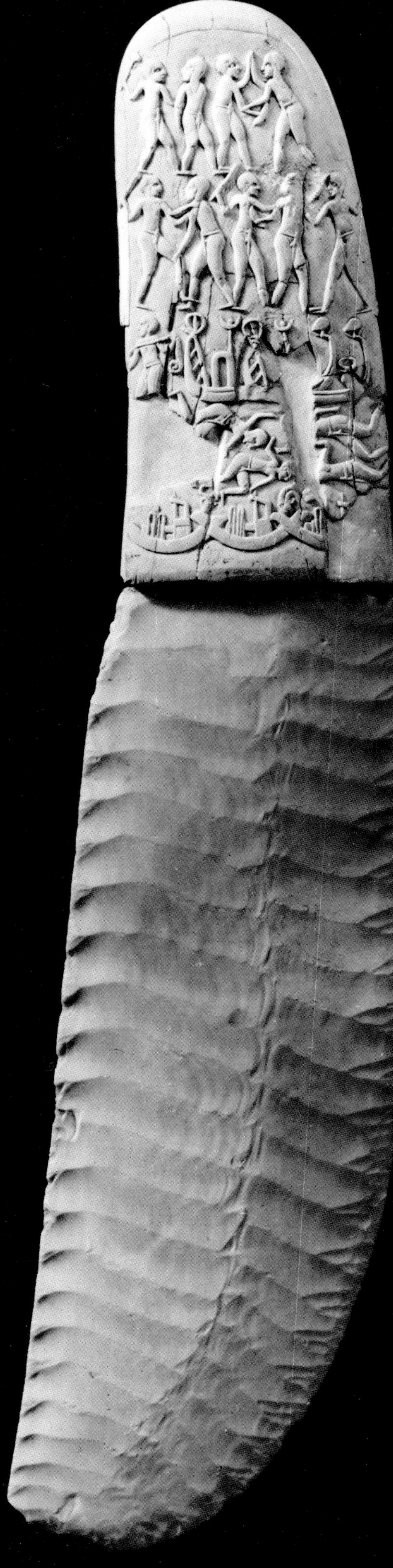

THE KNIFE AND ITS ORIGINS

"Possibly the first argument between hunters concerning the relative merits of cutting instruments took place a million or so years ago between two cavemen...with a couple of pieces of flint."
B.R. Hughes and Jack Lewis

Short history of the knife and the art of cutlery

The knife belongs to the great family of cutting instruments which is usually divided into three categories weapons, cutlery, and edged tools, to which should be added the blades of certain machine tools. Cutlery includes not only the knife as such, but also the razor, certain surgical instruments, scissors, various accessories such as the corkscrew, punch, sharpener, and others. But, on occasion, a cutler may be found making short-bladed weapons, like stilettos or daggers, and other makers encroaching on the cutler's domain. Until the 15th century cutlers proper were rare and knives were made by other craftsmen.

Be that as it may, the ancestor of the knife is a universal tool, product of early 'human' ingenuity. It goes back about two million years and was the work of Australopithecus. The tool in question was a simple flake of broken stone, its cutting edge having been created by the act of flaking it, and its virtue lying in its power to be the first extension of the human hand — or rather, perhaps, the mind. This weapon, first used for defensive and by turn for offensive purposes, definitively assured the dominance of the Hominidae over animals; thus, one may say that hunting created man. Later 'knives' were flaked on the other side of the slab, forming a bevelled edge. Around the year 50,000 B.C., this almond shaped, bevelled slab of stone evolved into one that was double-edged. Its uses were many: each side was a blade, the point an awl; in addition, the irregularity of the edge made it serve as a crude saw. The Châtelperron (central France) knife (early Paleolithic, c. 40,000 B.C.) is a true blade with a back. Even later, advanced ancient civilizations such as that of Egypt used knives made almost exclusively out of stone, although by this time they were polished (from 4,000 B.C. on and perhaps before). The materials used were generally flint, but also included obsidian (which is a kind of lava), ophite (a striped stone), or slate. Knives can also be found made from bone, ivory, stag-horn, and animal teeth. Probably even wood was used, although this fact would be hard to prove.

At an early date, man fashioned knives from copper, which was later abandoned because of its softness. In Europe, the Bronze Age began around 1500 B.C. The best bronze was an alloy containing 10 per cent tin. The first

The 'Jebel al-Arak' knife is one of the most beautiful objects of pre-dynastic Egypt (c. 3400 B.C.). The sculpted handle displays scenes of daily life 5,000 years ago. Musée du Louvre, Paris

This iron dagger with finely worked sheath is remarkably well preserved. It dates from the Hallstatt period (c. 600 B.C.) and was found in a tomb near Estavayer, Switzerland. Musée d'art et d'histoire, Fribourg

Two iron razors from a royal tomb in Westprignitz (today known as Seddin in East Germany). Preussischer Kulturbesitz, Berlin

The hunting knife on the left, with an iron blade, was discovered in a marsh near Bremen. It dates from the first half of the 14th century and measures 11 inches long. Museum für Kunst und Gewerbe, Hamburg

The table knife in the center is of Danish origin (14th or 15th century). The handle is silver and bears two mottos. Museum für Kunst und Gewerbe, Hamburg

This knife with a wide blade was used by a 'carving squire' of the middle of the 15th century. The engraved and gilded steel handle bears the arms of Philip the Good, Duke of Burgundy in silver with enamel inlay, the inscription says 'aultre n'arai' (I shall not have another one), and the sheath carries the collar of the Order of the Golden Fleece. The sheath is polychrome chased with silver. Musée de Dijon

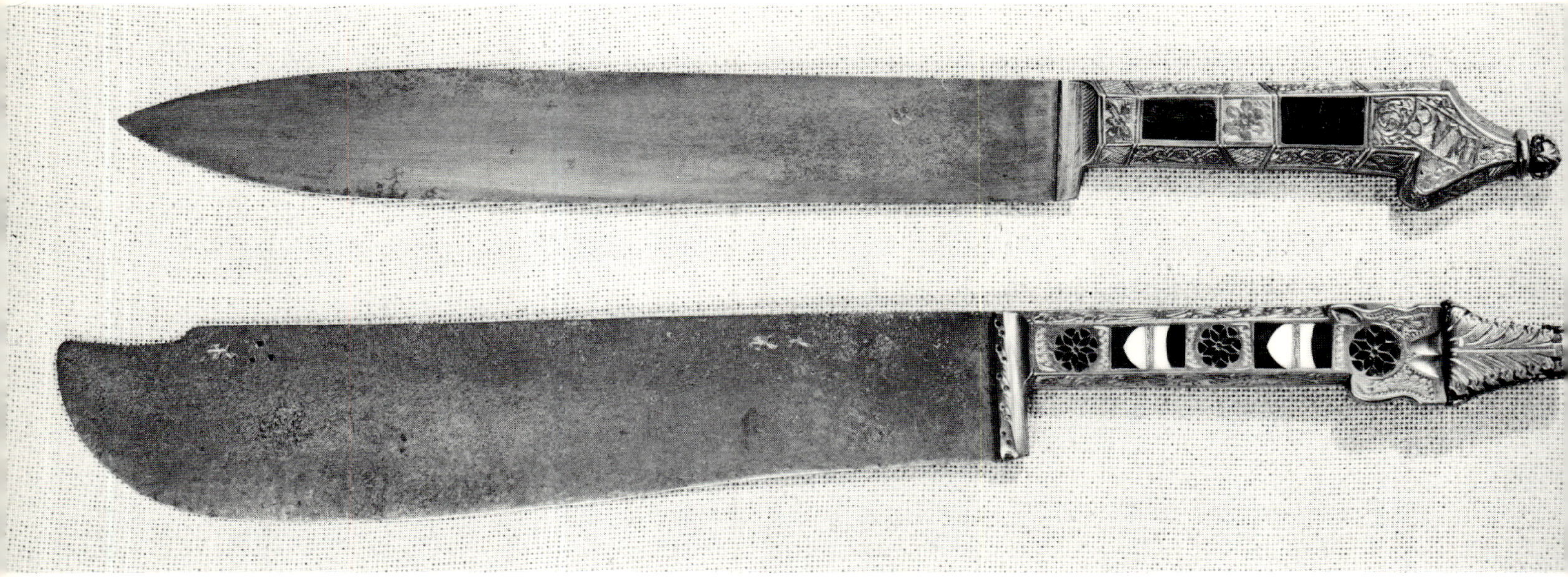

metal knives appeared shortly afterwards. They were cast and then hammered to harden the edge. Up until then, knives had been of one piece; thereafter most were made up of separate handles and blades. The most common method of joining the two was to insert the rearward extension of the blade, the tang, into the handle, or to join the handle at the base of the blade. Some knives were made with a socket. All of these articles, of fine work, were elegantly curved.

The slight swell of the spine was designed for increased force. The addition of the handle, of stag-horn, wood, or ivory gave the opportunity of decorating a utilitarian object. Working with metal, thinner, sharper, and stronger blades could be fashioned than were possible with stone. The form could be adapted to the function, and repair was made possible. In this period the first razors appeared.

The next stage, c. 1000 B.C., saw iron come on the scene.

Its mining posed more problems than producing bronze, which was originally obtained by direct reduction. As far as iron is concerned, a hole was dug into the side of a wind-swept hill, the bottom filled with wood, followed by a layer of ore. This was, in turn, covered with a second layer of wood and, when the wind was right, the bottom layer was ignited. The wind

A trenching knife and a presenting knife from northern Germany dating from the second half of the 15th century. The silver handles are set with precious stone and ivory. Germanisches Nationalmuseum, Nuremberg

The presenting knife, for serving slices of meat, was commonly used in the Middle Ages and into the Renaissance. This one has a brass carved handle and comes from northern Italy (15th century). Museum für Kunst und Gewerbe, Hamburg

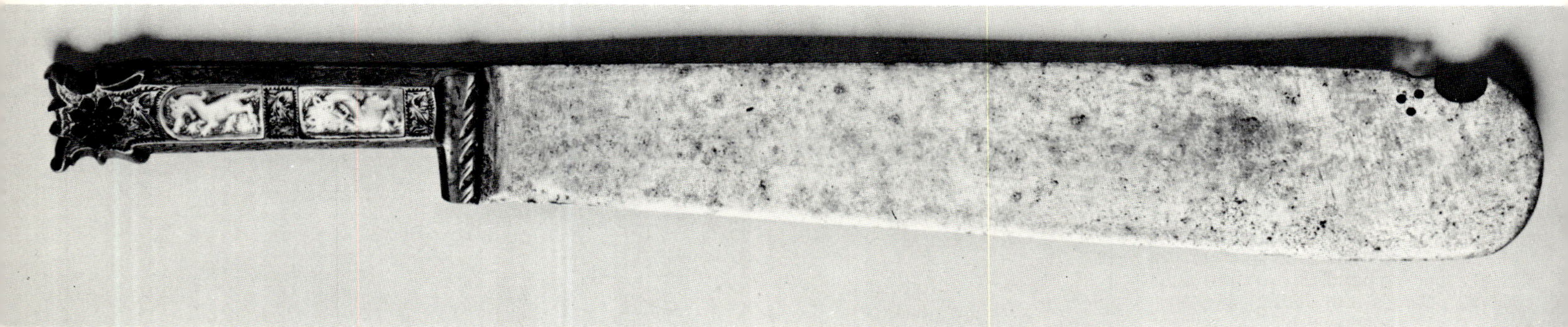

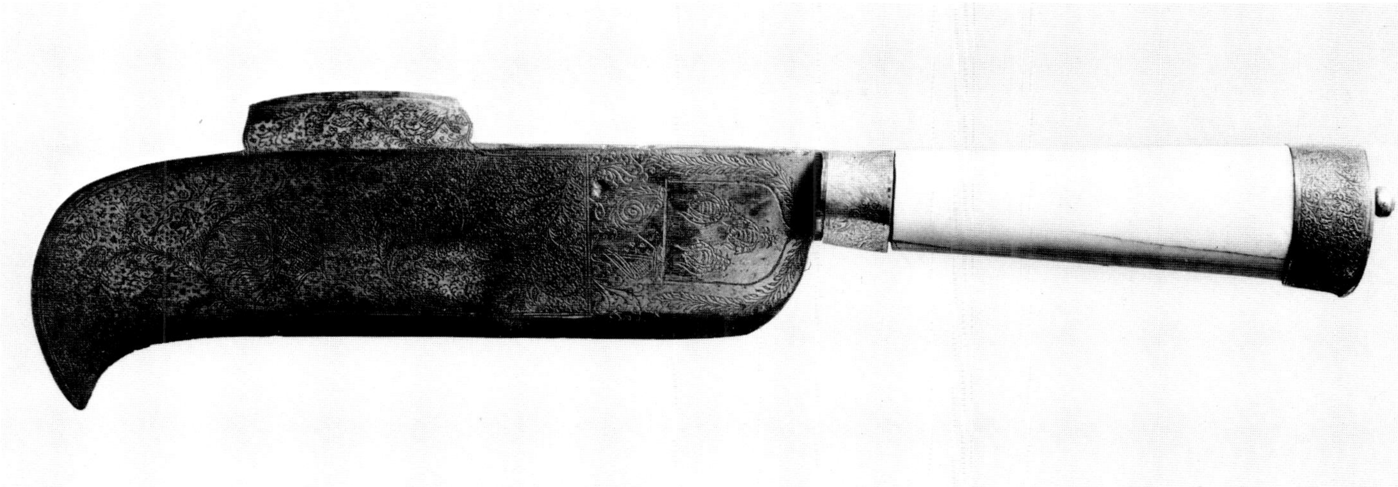

This disjointing-knife with the arms of Pontus de Thiard dates from the end of the 16th century. The handle is ivory and the butt and bolster band are engraved gilded copper. The blade is completely gilded and chased with decorations of foliage scrolls and hunting scenes. On the back is an inscription that says 'Maison de Tiar Bénissez le Seigneur' (House of Tiar, praise the Lord). Musée de Dijon

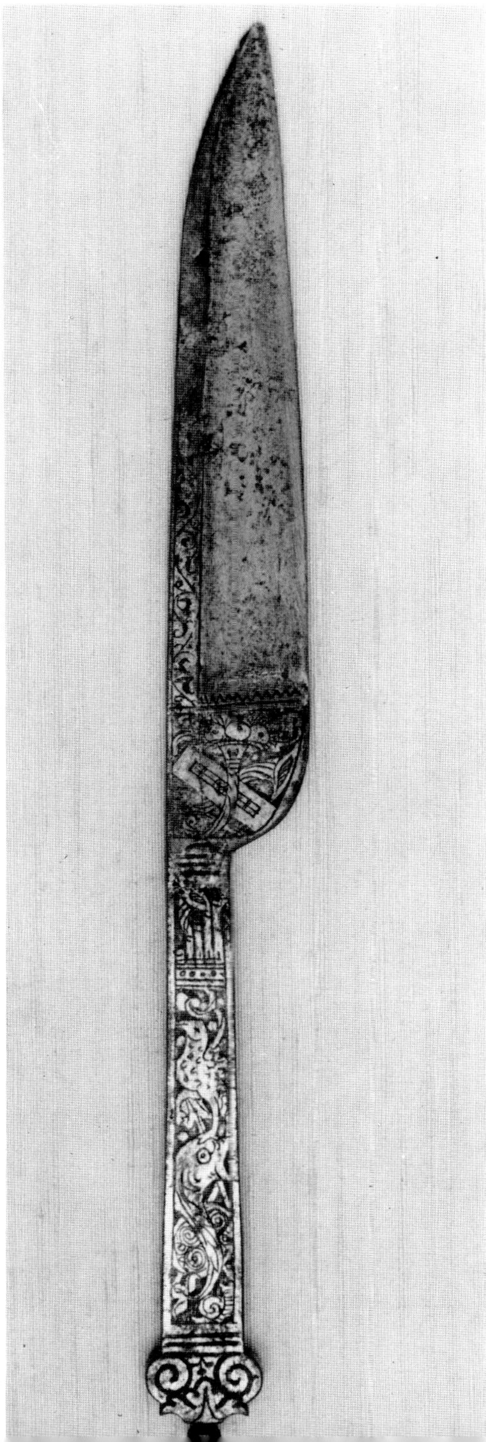

Left: The table knife is of Italian origin (beginning of the 16th century). The handle, which extends to the back of the blade, is decorated with arabesques and images of mythical animals. James de Rothschild Collection, Waddesdon Manor.

The French knife on the right is of an earlier period than the one on the left, even though it looks similar. The decoration is of foliage motifs and the handle is surmounted with an acorn. James de Rothschild Collection, Waddesdon Manor.

17th-century knife cases (1655 and 1654). Both bear the inscription 'I was in Sheffield made and many can witness I was not made by any man.' City Museum, Sheffield

This knife and fork have a double case. They are apparently of English origin and date from the 18th century. City Museum, Sheffield

Left: These forks and knives were made in Germany around 1650. The set on the right comes from Dantzig or Elbig. In both cases the handles are decorated with human figures, on the left of Biblical origin, and prosperous citizens on the right. The handles of the set on the left are enameled and set with precious stones; on the right simply carved and painted. Schatzkammer der Residenz, Munich; Schloss Nymphenburg, Munich

This travelling table set has a chain fastening system. The wood case is covered with leather and ornamented with silver plating. The scene shows Hygeia and Judith. The coat of arms could be that of the Jenatsch. Swiss, and probably dates from the end of the 16th century. Schweizerisches Nationalmuseum, Zurich

acted as a bellows, turning the wood into charcoal whose heat reduced the iron oxide ore to its metallic state. Once the fire was out, a few pieces of iron were withdrawn from the ashes. It was the discovery of iron which gave the greatest impetus to the progress of knifemaking by increasing the weight of knives, their strength, and consequently their ability to cut. Moreover, while iron was more difficult to work, it was more common, and therefore cheaper, than copper or tin.

Iron knives were made in most of Europe; but the credit for another invention, steel, belongs to India. There it was noted that when iron was heated with charcoal and cooled repeatedly, the carbon of the charcoal united with the iron, making it harder. The result was steel. Blades known as 'Damascus' first came from India. Steel was not used in Europe until the 10th century, when Toledo produced the best blades of the continent.

The Greeks called any instrument with one cutting edge *machoera*. Their barbers used a razor that did not have a handle. Butchers and sacrificers killed their animals with a single-edged bronze knife some 25 inches long. The same implement was used to kill the enemy in combat. It was made of one piece — the handle end having a ring for hanging. The Romans used several different kinds of knives, among them the *culter* or household knife, the *culter coquinarius* or kitchen knife, and the *culter venatorius* or hunting knife. They also had a whole series of other cutting instruments: fruit knives, toiletry knives, knives for fingernails, for leather, and for bleeding cattle — even folding knives. Knives used to perform sacrifices were highly specialized: knives for cutting the throat of the offering, flaying it, dressing it and for opening the entrails.

For cutting meat, the Gauls used a small knife carried in a sheath which hung from the side. For them, as for all 'barbarian' peoples, the knife was primarily a weapon for hunting and fighting. The *semispatium* (literally, half-sword), a kind of dirk made of bronze or iron, belongs to this category.

Besides the sword, the Franks had a large war-knife called *scramasax* which was shaped like a *butcher's knife*. In the Merovingian period this name was also used for a kind of single-bladed dagger. In the Middle Ages the knife was a double-edged weapon whose use was reserved to the commoners who were forbidden all other arms. It was suspended from a belt on the left side. During the Crusades, bandits and outlaws overran France. They carried a kind of *hunting knife* from which the *custile* was later derived, whence the name of *custiliers* given to foot soldiers of 14th and 15th centuries. In the time of the military-historical writer Blaise de Montluc (1502-1577), the *custell,* a kind of cutlass, was used to finish off the wounded on the battlefield, or to execute captives. Its exact shape is not known.

Although the *table knife* is mentioned in 11th century writings, it seems to have been little used until the 12th century and even after this period knives were rare. Only the rich had enough knives for all the guests at their tables — it was therefore necessary to carry a personal knife, kept in a sheath dangling from the waist or built into a sword scabbard. Women wore their knives in their garters.

While in ancient times cutlers were slaves, in the Middle Ages they were freemen who grouped themselves into guilds modelled after the monastic orders. The great center of French cutlery manufacture was Périgueux, but the best knives came from Toulouse. Paris in 1292 boasted only 24 cutlers who were divided between two guilds: 'the knife forgers' who forged the

This hunting knife is perhaps Hungarian despite its pseudo-Islamic style. James de Rothschild Collection, Waddesdon Manor

This pair of wedding knives dates from 1633. They were made by John Jenkes of London. The ivory and amber handle bears the name of the first owner, Anne Marse. City Museum, Sheffield

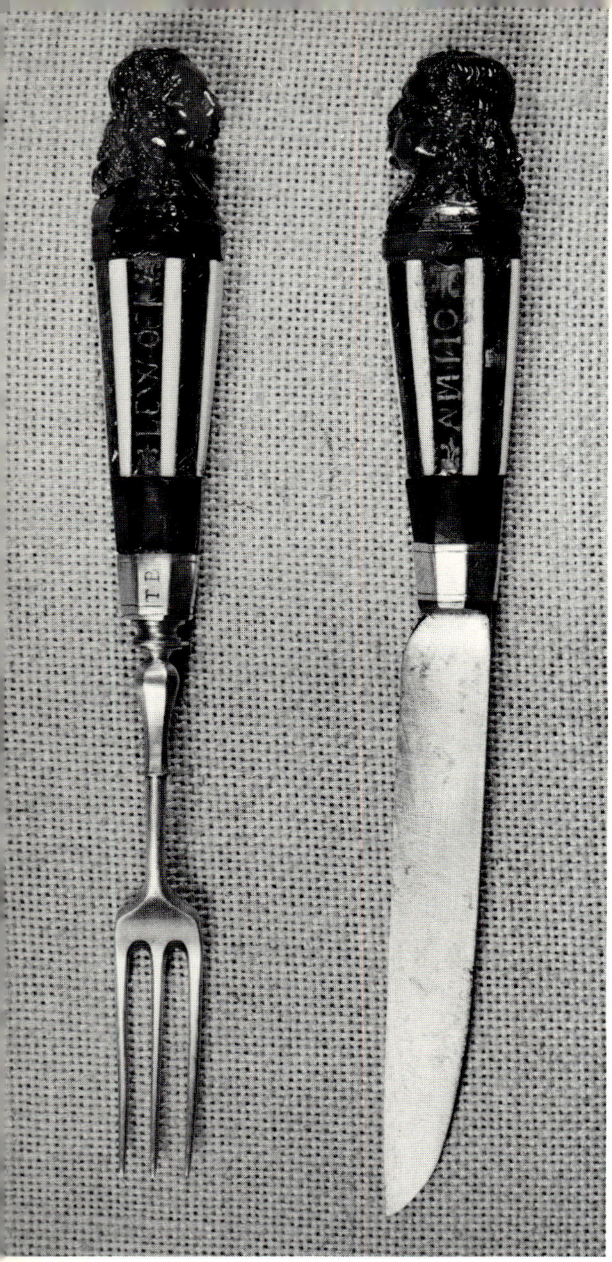

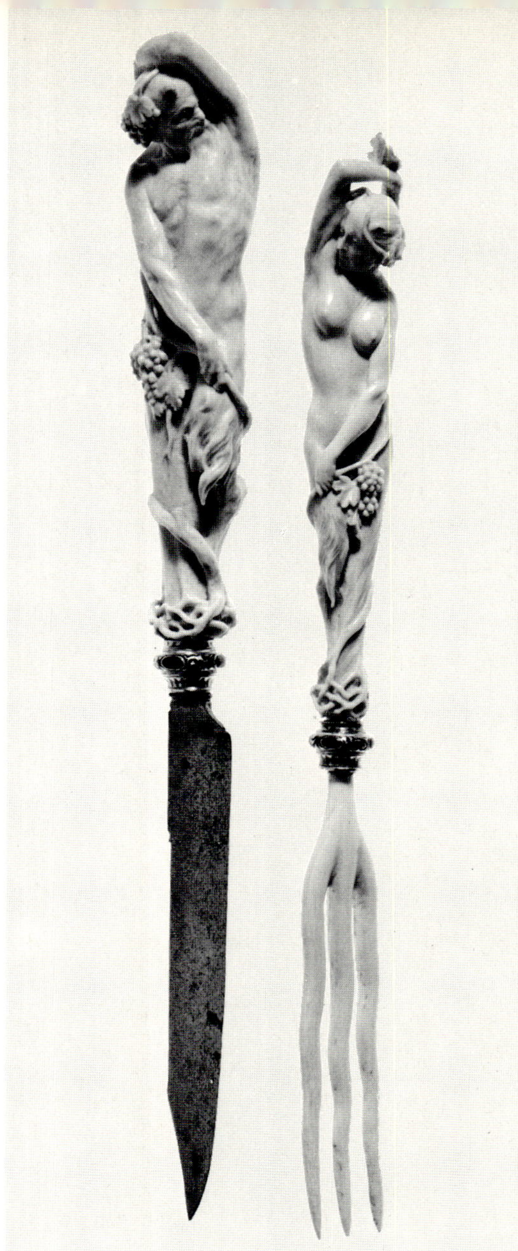

Left: The handles of this 1682 German table service are made of yellow amber. Germanisches Nationalmuseum, Nuremberg

The Bacchic figures and the fork are ivory. Flemish master, c. 1650. Germanisches Nationalmuseum, Nuremberg

Above: The carved handle has two nude figures with a dog at their feet. Netherlands, around 1700. Museum für Kunst und Gewerbe, Hamburg

Facing page, upper left: This Austrian sheath (about 1740) is ornamented with hunting scenes and contains a hunting knife as well as a smaller knife and fork. James de Rothschild Coll., Waddesdon Manor

Facing page, upper right: Tableware emblazoned with coat of arms. Kilian Fries, Zürich, around 1650. Schweizerisches Nationalmuseum, Zurich

Opposite: Knife and fork with twisted silver handles. Probably from Zoug, and dating from the latter half of the 18th century. Schweizerisches Nationalmuseum, Zurich

blades for knives and the handle-making cutlers. An increasing richness was lavished on the latter: the grips were made from gold, ivory, mother-of-pearl, ebony, and other exotic materials; some were studded or completely encrusted with pearls, coral, jasper, sapphires, turquoise, diamonds or rubies. In 1265 the Archbishop of Rouen forbade the nuns of Montivilliers to carry carved knives or those with work in precious metals. The collapse of silver prices after the discovery of the New World led to an expansion in silverwork used to decorate knife handles. Unfortunately, many of the most beautiful specimens were lost in the successive meltings down ordered by Louis XIV. But previous to this the art of cutlery had made remarkable progress because of the Crusades.

In the 16th century, the knives of Châtellerault were frequently considered the best in the world, even though the most luxurious knives were made in Paris. During the reign of Henri IV, elegant *table knives* were in vogue, with handles fashioned with bizarre figures, sometimes with china

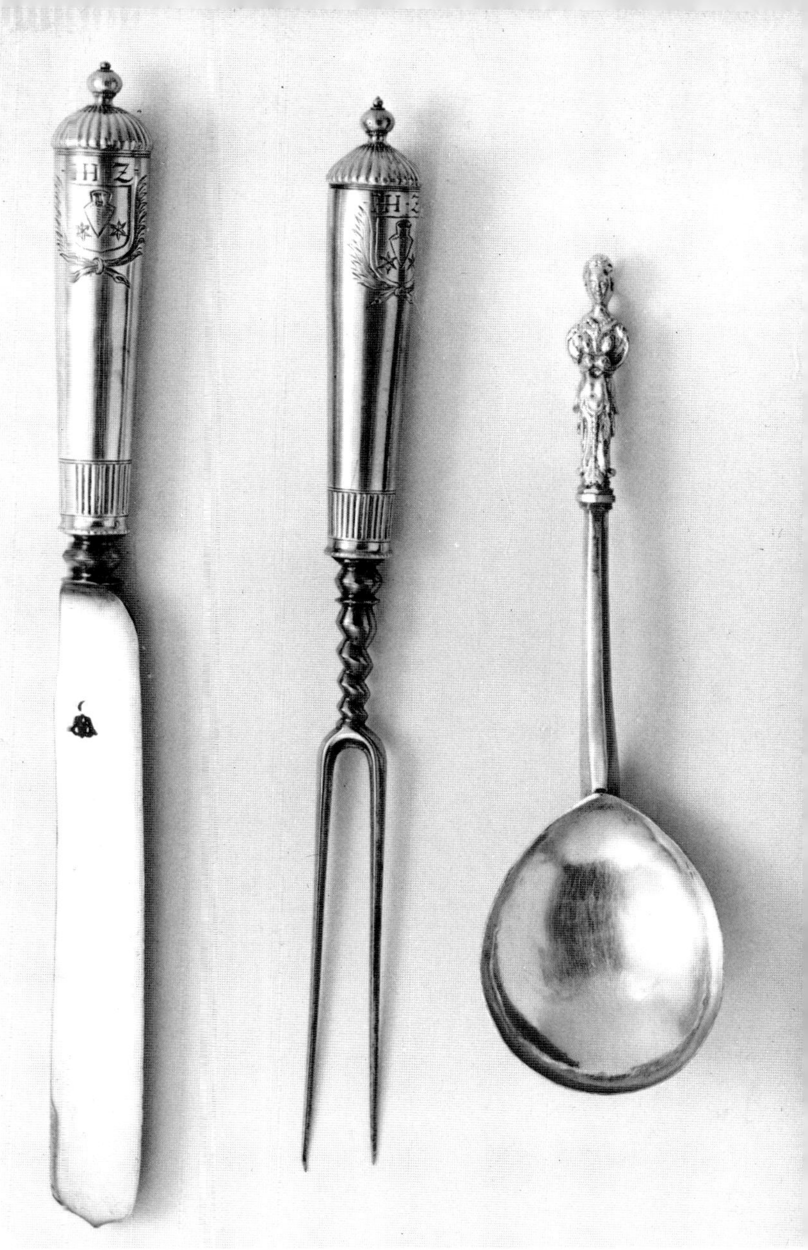

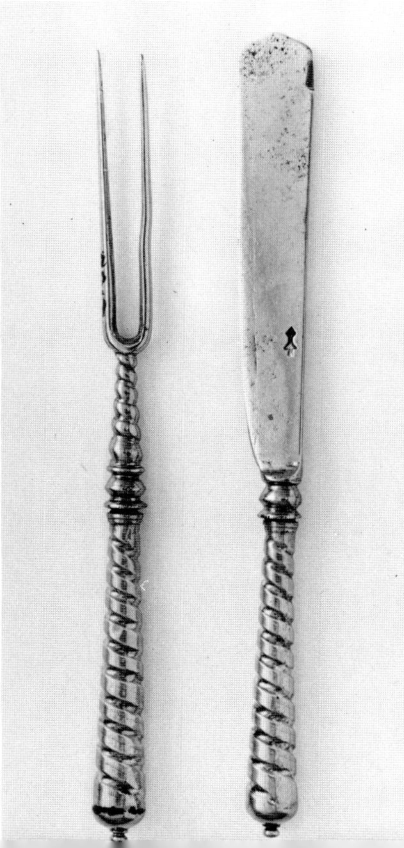

grotesques, which were known as 'china knives'. The greatest modification undergone by the *table knife* was the rounding-off of the blade-end. Until the end of the 16th century, the absence of a fork meant that the fingers and the knife were the tools used to bring food to the mouth. The knife had to be pointed in order to spear the meat. After the adoption of table forks the need for the point disappeared. The memorialist Tallement de Réaux (1719-1790), writes that it was Cardinal Richelieu who brought about this development. He was obliged to dine next to the Chancellor Séguier who ate in a coarse manner and picked his teeth with his knife. This annoyed Richelieu so much that he ordered his steward to round off the points of his knives. In 1669, an edict issued at Lille prohibited "every person of whatever quality to bear pointed knives... with the injunction to round off what knives they possessed."

Chaucer, writing in the middle of the 14th century, used that old English phrase: "A Scheffield thwitel baar he in his hose." This proves two things: that Sheffield was already renowned for its cutlery, and that an all-purpose knife called the *thwitel* was common at the time.

In the 16th century, French knives were for the most part made in the cities of Moulins and Langres. Cutlers did not shrink from making the rounds of country inns themselves to offer their work for sale. The use of *folding knives* became widespread at this time as citizens of the towns kept the habit of carrying a knife when they went out to dine.

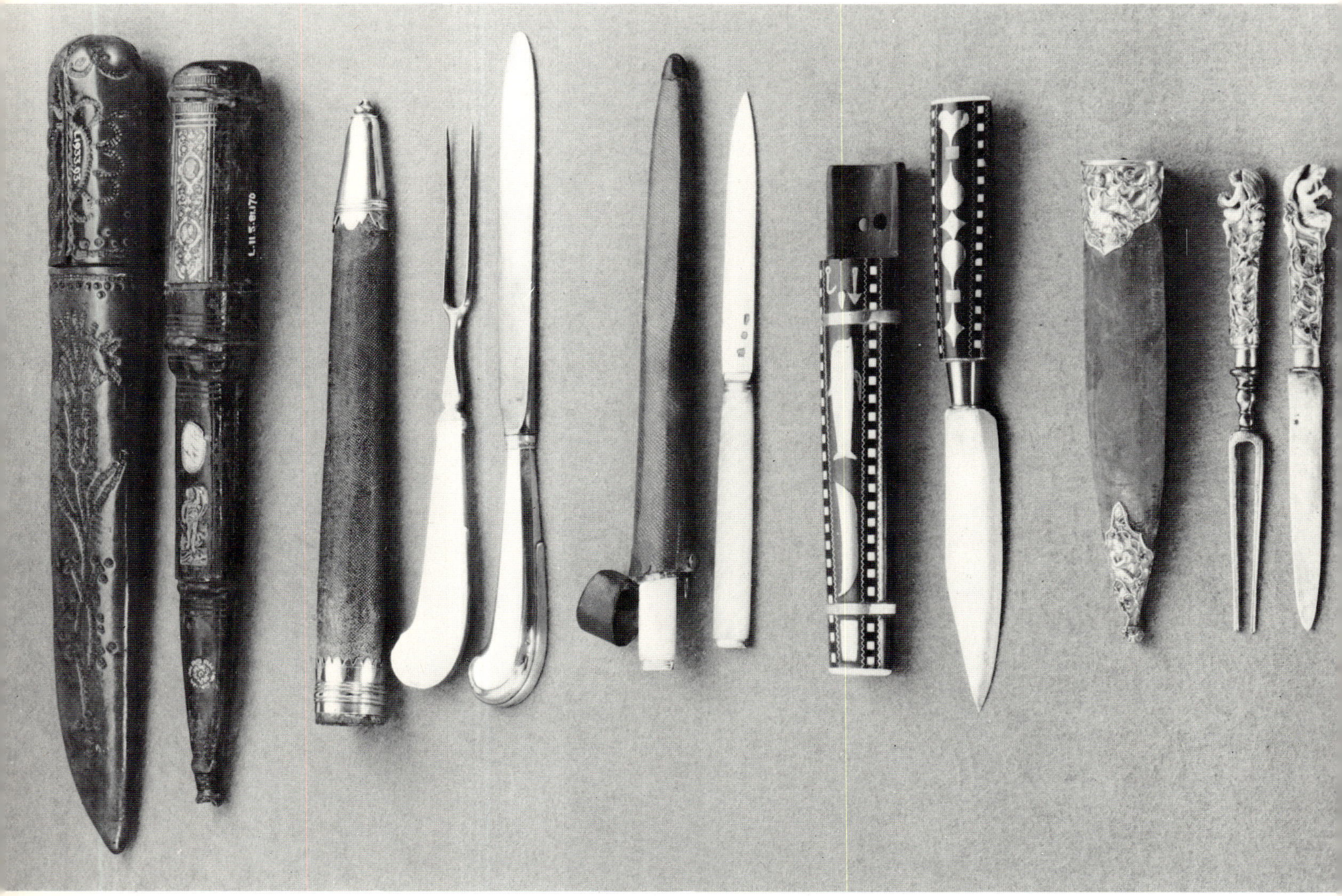

This collection of knives, each with its sheath, covers three centuries. The sheath of the first (from left to right), dated 1779 and of Dutch origin, is leather, decorated with motifs of flowers and birds. The second sheath is gilded; it is French work of the 18th century. The knife and fork next to it were made in London (middle of the 18th century) and fit into a wooden sheath covered with black shagreen with a silver chape. It was a travelling set. The two following pieces, one of which is in its shagreen case, are French dessert knives from the second half of the 18th century. The two black and white pieces are Danish. The handle and case are made of hardwood inlaid with brass, whalebone and pearls; they were used by dolphin fishermen in the Faroe Islands a hundred years ago. The last three pieces are Dutch (c. 1675); the sheath is red velvet with gold chapes. The chapes and the handle are ornamented with chased Bacchic scenes. City Museum, Sheffield

The anatomy and technology of the knife

There are many different kinds of knives. But the typical modern knife can be divided into four parts: the blade, the tang or extension of the blade to which the handle is fitted, the handle itself, and the hilt. If the hilt extends out or has *quillions* to keep the hand from sliding down on the blade (or in fighting knives, the adversary's knife from sliding up) it is more properly called a guard. The analogous part in a *kitchen knife*, and some other knives, is called a bolster and is frequently a metal ring which reinforces the joint between the blade and the handle. In *folding knives* the mount or fitting at one or both ends of the knife case is also called a bolster. The blade itself is divided into various parts: the back or spine which supports the edge and narrows as it runs from the hilt to the end of the blade, the end, which can be round, blunt or pointed, and finally, the cutting edge.

The successive steps of knife manufacture in the 17th and 18th centuries were described by the specialist Gustave Saint-Joanny a hundred years ago: "First the iron and steel is given to the *stamper*; his instrument is the *drop-stamp* or *tilt-hammer* which rises and falls by means of a waterwheel outside in the river. It can strike 200 to 500 blows a minute. The *stamper* is followed by the *blacksmith* who, with his weighty hammer raised by the strength of his arm, strikes the steel anew to give it the shape of the blade, the plate, or the spring. After the blacksmith, come the *filer*, the *temperer*, the *grinder* and the *furbisher*. While these craftsmen fashion the parts that will make up the knife, blades, plates, and springs, others prepare the handle. The *sawyers* of bone, of stag-horn, of boxwood, of ebony, cut these materials; the *bleacher* dries the bone hafts on his roof; the *straightener* dresses those which are of stag-horn. When all the parts of the knife are brought together the master gives them to the *assembler*, whence the knife passes to the *honer*."

Since the 19th century the manufacture of knives has become more systematic. From forty to sixty operations may be involved, but several characteristic steps can be distinguished. First we will consider the stages in the making of a blade, the other procedures, and the assembly.

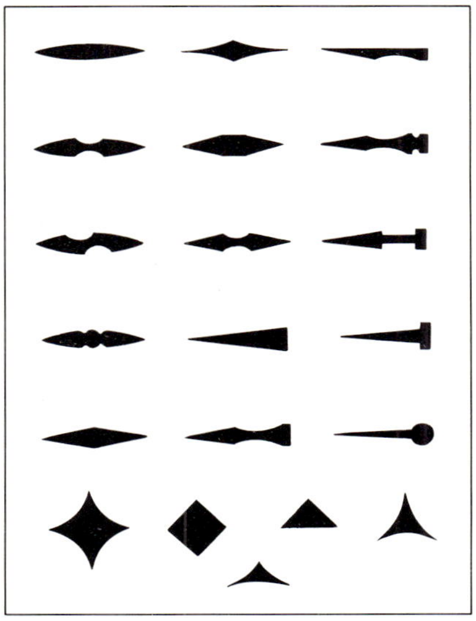

Blade sections can be completely different from one cutting tool to another.

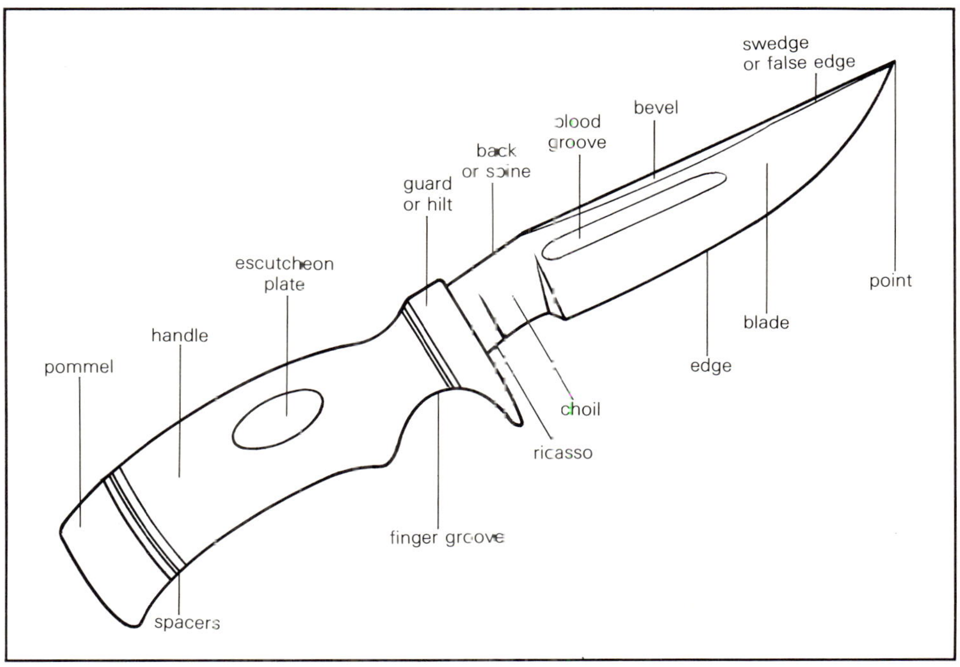

Essentials of a typical sheath knife

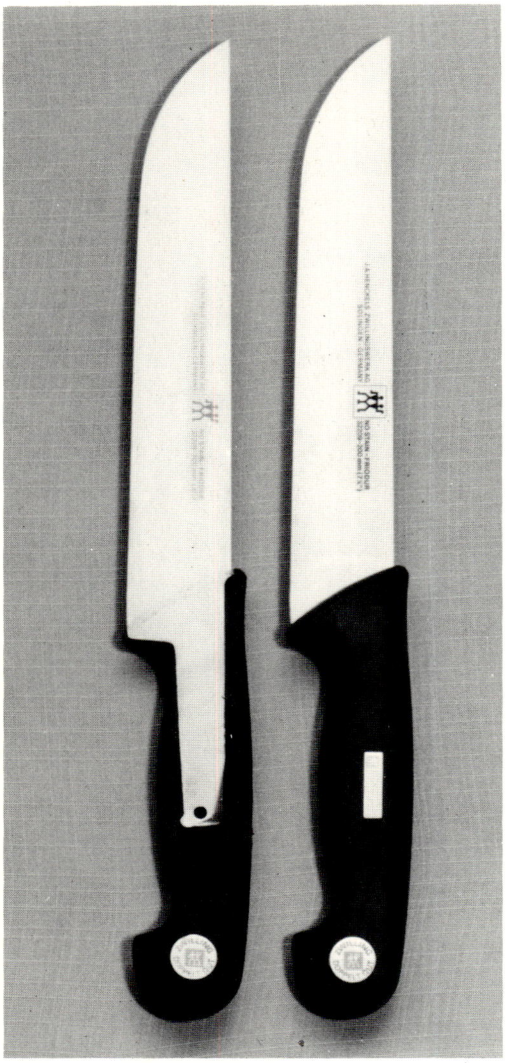

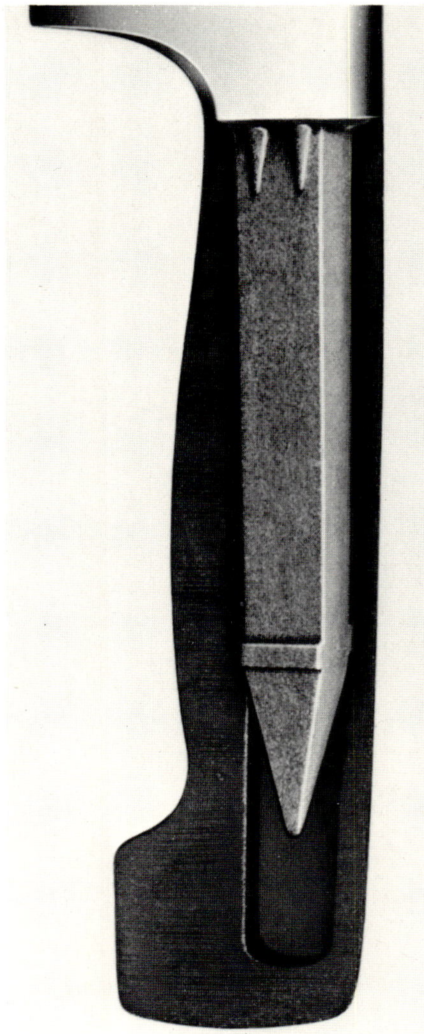

These two photographs illustrate two kinds of tang, with and without a rivet. The second one shows how the tang fits into the handle; the protrusions of the tang fit tightly into the handle.

Some cutler's marks

Below: the left hand diagram shows how a blade, its curve, and the length of the whole weapon are measured. The right hand diagram shows several different kinds of blades: a back curved blade with a back-edge point; a straight blade with a back-edge point; a back curved blade with a centered point; a straight blade with a centered point.

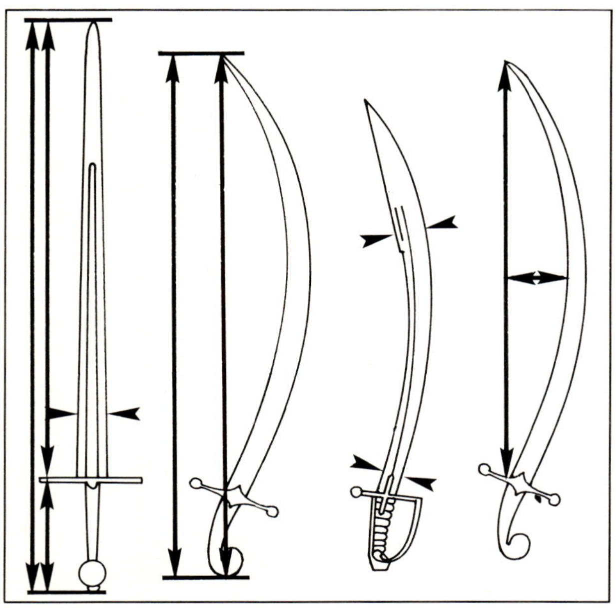

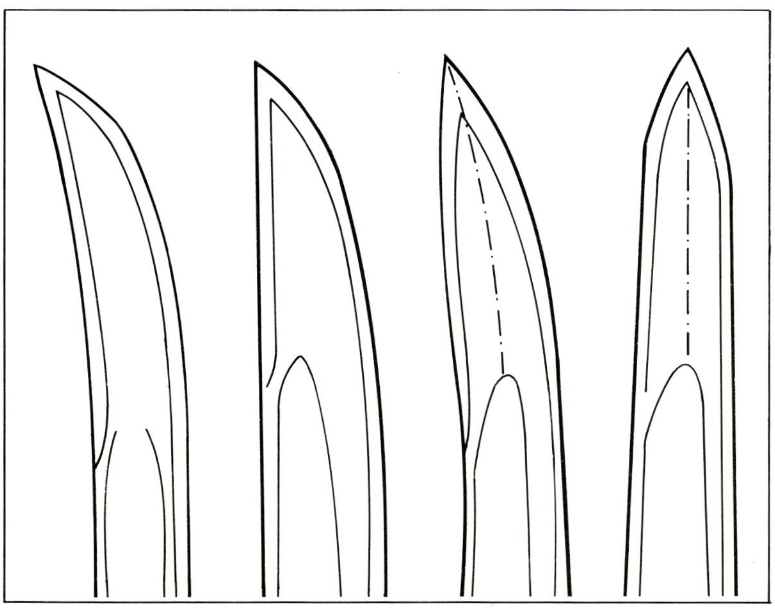

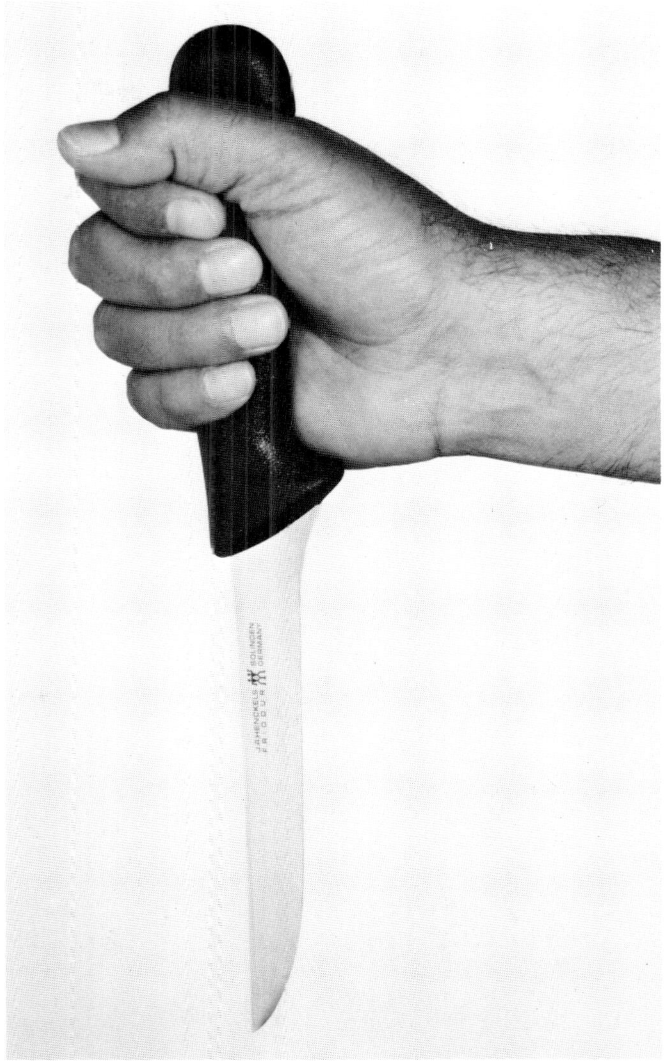

The merits of a handle with a protruding hilt are considerable: the user obtains a good grip on the knife and the danger of an accident is reduced to a minimum. Such a handle is particularly recommended for kitchen and butcher's knives.

To resharpen an edge to a knife is an art, but an art which is relatively simple when compared to sharpening. It is enough to pass the length of the blade three or four times on each side over a butcher's steel.

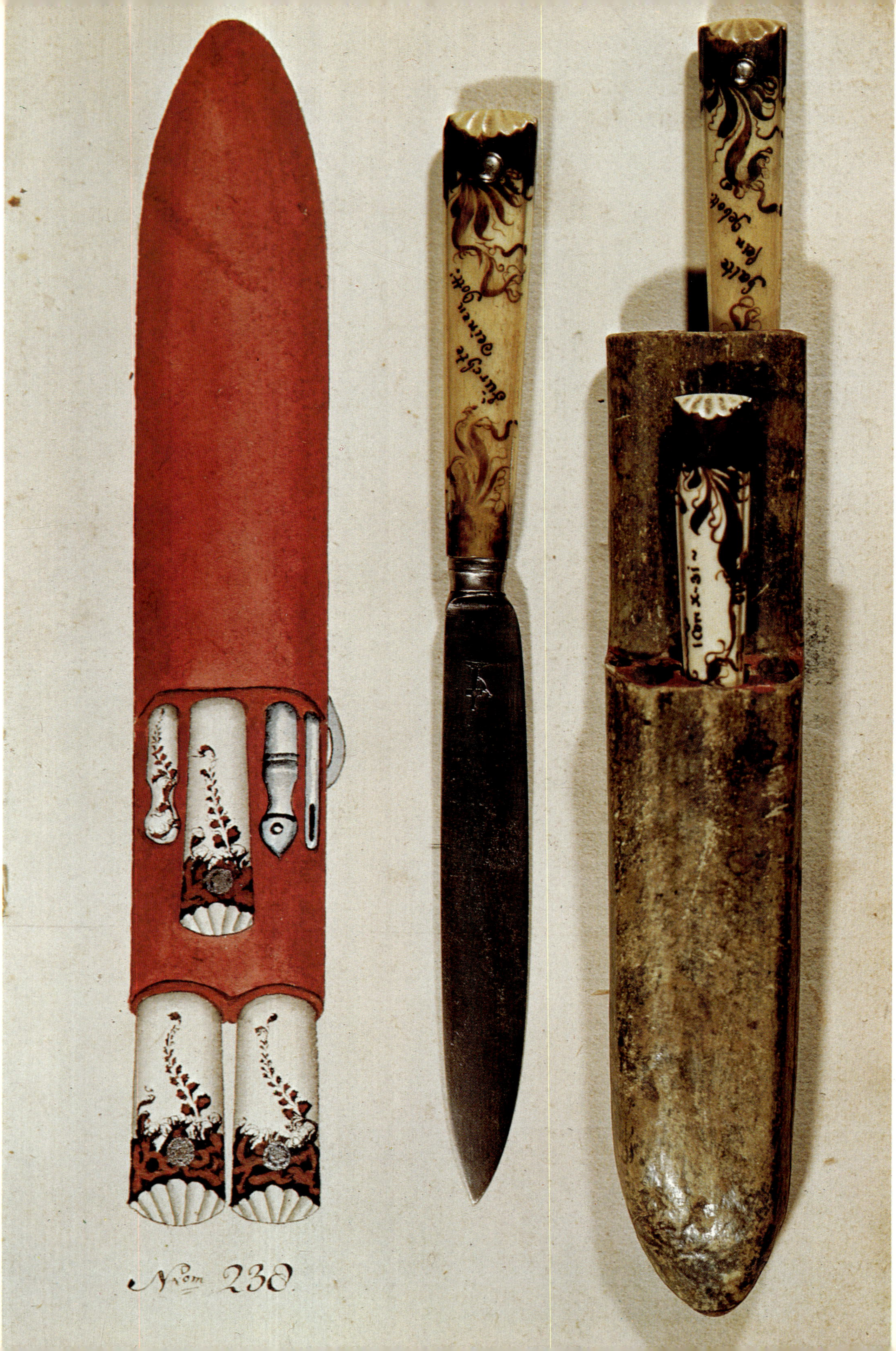

The business card of an 18th century French cutler. City Museum, Sheffield.

Placed next to an illustration from an old German sample catalogue of the 18th or 19th century are the actual knives illustrated. The sheath can contain six pieces, of which three are knives. On the handle of one is the inscription 'Fear God', on the other 'Respect His Laws'. Deutsches Klingenmuseum, Solingen, Germany

The first step in making a blade is the forging, an operation which consists of cutting bar metal into blanks which are drawn out by small power hammers to form the blade and the tang. The desired form is shaped with shears and finished off by a milling cutter. Annealing (see glossary at the end of the book) precedes the bleaching and the first grinding, which evens the thickness of the edge; the punching is done by a press and the blade then passes on to a craftsmen who corrects the irregularities. Tempering used to be the most delicate operation. Its specific nature varies according to the quality of the knife, but the procedure consists in suddenly cooling the heated metal (quenching) to increase its hardness. Tempered steel is hard, but brittle. To remedy this it undergoes a 'reheating' or 'drawing of the temper', which is a reheating but to a temperature below that of fusion. During the trimming, a craftsman restores to true the distortion of the blade caused by the tempering. The last steps are the sharpening, followed by a polishing which gives the necessary finish to the blade. At the same time the handle follows its course to completion through the steps of being cut, prepared, fashioned, drilled and polished; and so also the ring or the hilt. Assembly consists of fixing the hilt to the handle, fitting and cementing the tang, the polishing of the hilt and the handle and the final sharpening. The different manufacturing procedures will be discussed in the third chapter.

In the beginning, only carbon steel was used, that is, a mixture of iron and carbon — a margin of 1 to 1.5 per cent of impurities is ignored. Until the 18th century a partial fusion of carbon and iron was achieved by strewing charcoal on red-hot iron bars which were then hammered. Another method was to stack the metal with alternate layers of charcoal and to place more charcoal all around the pile; this mass was then heated for hours, producing wrought steel. This, like smelted steel (made by melting iron and carbonizing it in the liquid state) was completely suitable for making quality cutlery. The carbon content of the hard steel used in cutlery, with the exception of some surgical instruments, was of the order of 0.6 to 0.7 per cent.

In the 19th century the development of blast furnaces helped to bring large quantities of cast iron to the market at a low price. Various processes of steel-making were patented: Bessemer (1856), Siemens-Martin (1864) and Thomas-Gilchrist (1876). Stainless steel was discovered between 1903 and 1912 and led to decreased use of carbon steel. There are several types of 'stainless' or corrosion-resistant steels, the most common being sometimes known as '18/8' which contains, besides carbon, 18 per cent chromium and 8 per cent nickel. Other elements can be used in steel, including vanadium, tungsten, molybdenum. These discoveries in steel production led to an increasing mechanization in the manufacture of cutlery, a revolution in the procedures of forging — and the ruin of many craftsmen. At present, the largest cutlery producers are Great Britain (at Sheffield, London and Birmingham) and France — each with 20 per cent of the world's total production — and Germany (at Solingen) which alone is equal to both of them. The combined tonnage of the other producers — Sweden, Italy, the United States, Belgium (whose noted centers are Liège and Namur) and Switzerland — does not approach that of any one of the others. In the United States, manufacturers are concentrated in the New England states, but there are also a few centers on the Pacific coast, as well as in the central states. Japan works more and more for certain American firms. But the pride of the United States lies especially in its handmade knives.

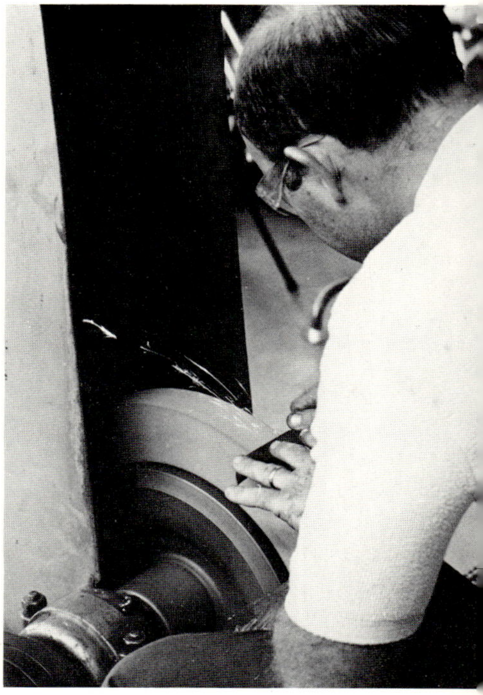

Lebanese cutlery has been renowned for thousands of years, as the discovery of a silver knife inlaid with gold in a royal tomb of Byblos (19th century B.C.) demonstrates. Today this trade still keeps its craft aspect; the most famous center is in Jezzine, 50 miles from Beirut. Only the materials have changed over the last century — apart from copper, silvered and stainless steels are used. Lebanese Tourist Office, Beirut

Handmade cutlery has in recent years seen a new resurgence of interest, thanks as much to collectors as to campers, hunters and fishermen. Above right, an American cutler at his grinding wheel. Centofante, Tampa, Florida.

Below: These grinders of Thiers, France, lying flat above their grinding wheels carry on an age-old method. This system, picturesque and skilled as it may be, has almost disappeared except for a few family enterprises. Editions Réunies, Panazol

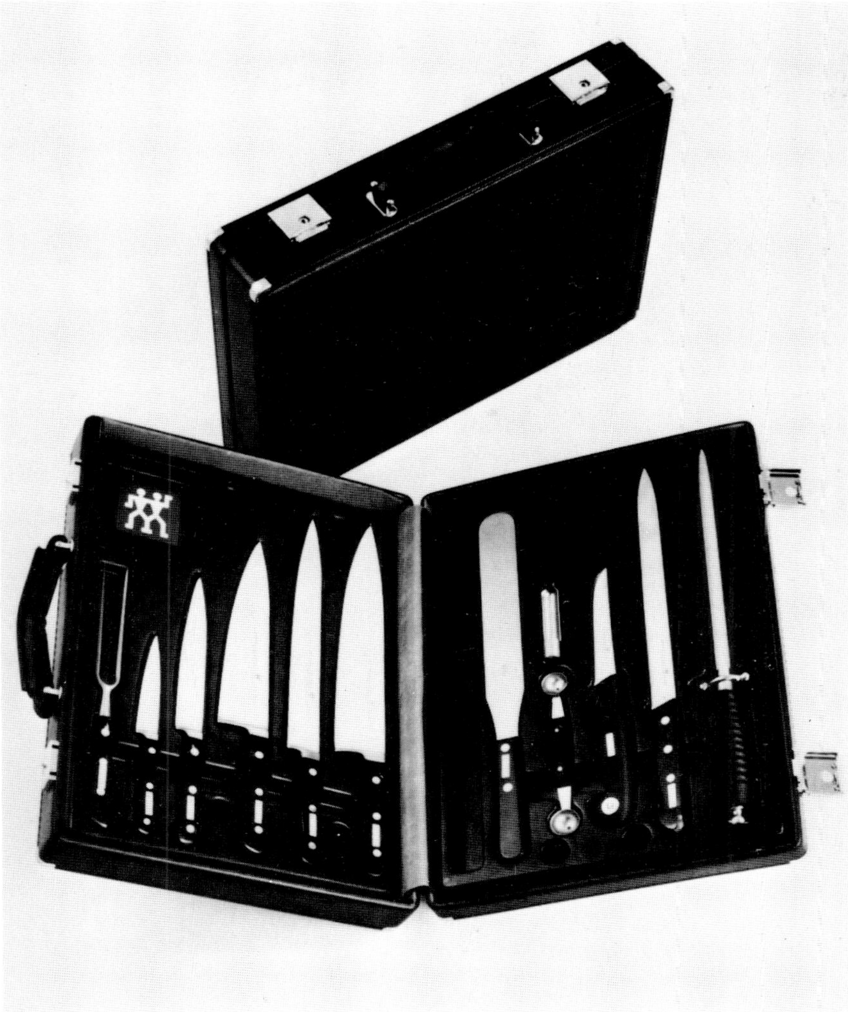

The making of sheaths and cases is complementary to the knife industry. Here is a professional case for a set of kitchen knives comprised of carving fork, skewer and butcher's steel. Zwilling, Solingen

This 1901 advertisement shows the assortment offered by Thomas Turner, merchant-cutler of Sheffield. It contains ninety pieces, scissors, razors, and an impressive panoply of folding, grafting, kitchen and carving knives, each marked with its price. City Museum, Sheffield

THE OUTDOOR KNIFE

"In many ways political history was carved with a blade..."

Norman M. Strung

Hunting knives

The first arms, whether hand held or thrown, were hunting weapons. It was not until later that they were turned by man against man. The middle Paleolithic age saw the rough prototypes of daggers fashioned from fragments of bone and horn. Later, the blade became longer and more narrow, as it was adapted to the needs of hunting and the dressing of furbearing animals. The hunter's armory grew rapidly (the bow and arrow, boomerang, sling, javelin, etc.) but it is mainly through the use of the lasso or of nets and traps that the knife came into its own, being the instrument with which the ensnared animal is finished off and then skinned. The *machoera* of the Greeks of Homeric times, worn alongside the sword, was used in hunting as often as for sacrificing or for the kitchen. The ancient Persians and Assyrians hunted lions not only with the bow, but also with the knife. The famous Roman short sword was at the same time a weapon for war and for hunting. Up until the 16th century the distinction between the two was arbitrary.

The domestication of the dog was followed by that of the horse, and this opened the way for *riding to hounds*. This type of hunting began in ancient Gaul: the preferred quarry of the Gauls and early Franks being the tasty aurochs or wild ox. In the 12th century, a sort of cutlass, which was hung at the belt next to the sword, was used to finish off the animal as well as to dress and skin it. In this period a *hunting knife* was any short blade fixed to a handle carried by the huntsmen, the whippers-in or the beaters. Its form was identical or similar to that of present day *butchers' knives*. The true *hunting knife* did not appear until the end of the 17th century. Until then, the hunter took the field with his sword and used it to strike down the animal. As it was a heavy weapon, a dull part of the blade was grasped in the left hand which directed the blow, while the right grasped the hilt and pushed in the sword in the normal manner. If the blade had a cutting edge along both sides, the hunter protected his left hand by wrapping the blade with leather or cloth. The sword scabbard held a smaller *skinning knife* called a *bastardeau*, and a *scraper* was used to detach the meat from the bone when dressing large animals. Bear hunting differed from other kinds of hunting by its special technique, in which the animal was forced to stand erect by the use of pikes and was then stabbed.

When hunting with a pack of hounds clearly became a distinct kind of hunting, and when a *court sword* became a ceremonial arm that would break

The gilded brass handle of this hunting knife dates from about 1700. The blade, however, is older. The copper scabbard has gilded brass chapes, and is decorated with hunting scenes. All the pieces are of German workmanship. Germanisches Nationalmuseum, Nuremberg

when used against a boar's head — then the *hunter's knife* came into being. Its purpose was to deliver the coup-de-grâce to the animal, that had been drawn down by the hounds. But we are getting ahead of the story. A few words should be said about *hunting to hounds*. A princely sport for hundreds of years, in its current form it originated in France and was brought to England by the Normans. Originally only the stag was hunted — other game not being considered worthy — until fox hunting eventually became pre-eminent and in this form the hunt established itself in America. The conventions of hunting to hounds became fixed and its 'courtly' rules generally accepted around the year 1580. The hounds are said to be 'running hounds' and those with a particularly acute sense of smell, 'tufters'. The general term 'hunting at force' comes from the latter part of the chase, when the animal is near death and turns at bay. The mort is blown and, once the animal is dead, comes the *curée*, when a hunt servant removes the hide upon which the quarry or parts of the prey fed to the hounds is placed.

The modern *hunting knife* is a short sword with a wide blade, which the huntsman wears on his side fixed to a belt. His hands are thus free, permitting him to dismount and to finish the animal without hindrance. 'Courtly' venery no longer permits the stabbing of the fleeing animal with a sword from horseback, as was done in the Middle Ages and the Renaissance, and the *hunting knife* comes into play only when the animal stands at bay.

The *disjointing knives*, *choppers* and *hatchets* used to finish and prepare the quarry in earlier times have been succeeded by a single strong knife. Moreover, the contemporary practice of finishing off the animal with a gun has reduced the knife to playing a decorative rôle. As for the so-called *hunting bayonet* fixed to a rifle and the *spear-sword*, these fall in the category of shafted weapons rather than knives.

Unlike the preceding cutlers' marks, these were used by makers of side arms. Sometimes two marks were placed on the same blade: that of the city or the guild, and that of the manufacturer. In addition there may be good luck marks, the coat-of-arms of the owner, inscriptions, etc.

The arms of war

Throughout the centuries it has been very difficult to distinguish between implements of the hunt and weapons of war. The outward form of a weapon often has little to do with its final purpose. Any sharp or blunt instrument can play the rôle of a weapon and become the handiest means of attack or defence against others — be they men or wild animals.

The first arms were those of hand-to-hand combat. In this way they assisted the process of natural selection, since their efficacy relied solely upon the user's skill and muscular strength — hence their generic name of *hand arms*. Two early kinds of arms can be distinguished: the various types of knives, and the striking weapons. The latter were limited to various cudgels, bludgeons and maces.

Soon man sought to strike his adversary while at the same time being protected from his opponent's blows. This he could do in two ways, either by using defensive weapons such as shields, helmets, armor and so forth, or by extending the distance between himself and his adversary by lengthening the blade of his dagger — which led to the birth of the sword. Another means of extending this distance consisted of mounting the blade on a long stick, resulting in shafted weapons such as the lance, the pike, the spear and so on. He also employed throwing weapons in the form of javelins, arrows shot by bows, and all other throwing or missile weapons. The variety of arms multiplied rapidly. After the sword and the rapier, other hand arms came in-

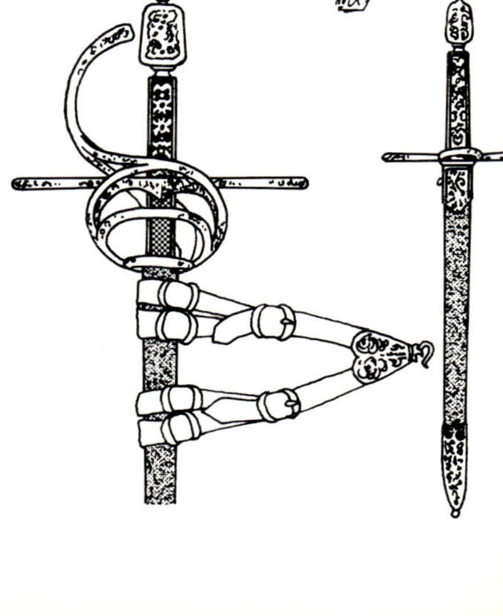

to being, such as the sabre and the bayonet. An interesting point is that in French this whole category of arms is called 'white weapons' meaning 'shining', referring to the glitter of light when an edged weapon was drawn from its scabbard.

Among defensive weapons, armor made its appearance — and in answer to it were born the mace and the ball and chain. The halbard and fauchard were new additions among shafted weapons. But naturally, with the increased distance between adversaries came a decrease in courtly valor. The feat of arms, at one time considered a revealer of merit (*virtus*) on the part of the combatant, became simply a means of destruction. The use of fortifications and war machines further enhanced this aspect. The medieval church tried to counter this mechanization and dehumanization, first by prohibiting the use of certain weapons, and then when that failed, by promoting a new idea: Chivalry and the defence of Christendom. To bear an arm became a mark of confidence on the part of the sovereign, the bishop or the lord. Courtly arms to be used at jousts, parades, in town and at court came into being, and at the same time, purely honorific arms, which could not be used upon the field of battle, appeared as a reward for merit. It was but one step from that to an aesthetic conception of arms. The introduction of gun powder, without which, as the historian Monluc says, — "So many brave and valiant men would not have died at the hands of cowards" — marked the end and final development of classical missile weapons, the beginning of firearms, and the final decline of hand weapons. In 1588 the defeat of the Spanish Armada brought final proof: even at sea the sword could not compete with gunpowder. At the very moment when the *hunting knife* was born, that of war ceased to be anything more than an auxiliary instrument, except the *cavalry sabre* and the lance, which lasted until the 19th century. Only the introduction of the *bayonet-sword* at the end of the 17th century, and some practices of modern guerilla warfare, are exceptions to this generalization. The form of the neolithic dagger was so ideal that during the centuries only the materials from which it was made were changed. The process of forging a less breakable metal made it possible to lengthen the blade. This tendency grew due to the skills of Iberian craftsmen, followed by German ones who invented a very strong steel. A whole series of large swords came into being, with very sharp edges and above all great flexibility. Some were simply gigantic — the length of a man — and reserved for use by the strongest. The thrust — with an *estoc*, a small *stabbing-sword*, worn at the girdle — was succeeded by the cut and thrust, that is, the use of the edge and the point. Another term used in the same epoch, *breech knives*, actually refers to shaft weapons mounted on pikes.

But in the meantime what had happened to the knife, or rather the short bladed hand arm, since the dividing line between it and the rapier or short sword would be hard to draw? It had not, of course, disappeared. In Roman times, Thracian gladiators used the *sica* or *curved knife*, which they handled with the blade turned inwards because it was used as a disemboweling weapon with upward cuts — a technique still used today in certain parts of the Mediterranean. In the Middle Ages, a sort of single-edged dagger called *arming-dagger* was used as well as the sword. The name *brackmard* was applied to a weapon with a straight, flat, wide blade, very pointed and sharp at both edges. *Hook-knives* and *plate-knives* were weapons with a thin blade made to slip between plates of body armor in order to finish off the wound-

This dagger of the 'cinquedea' type comes from northern Italy. It dates perhaps from the beginning of the 16th century, but could be as old as the end of the 15th.

Towards the end of the 16th century, the 'rapier-dagger' set replaced the 'estocade-town buckler' set. The reason for using a dagger, often called 'left-hander', is to avoid the thrust of the adversary's sword.

ed, or simply to kill the armored horseman who had fallen off his mount. They were also used to bring down horses. From the 14th century a strong dagger with a reinforced point came into use. Called a *chain-mail piercer*, it could go through a shirt of chain-mail armor.

As we have seen, the dagger is one of the most widely used hand arms. It appears in Italy in the 14th century as a long double-edged knife 10 to 20 inches long with a narrow, point blade. Civilians wore it hanging from the belt, whereas soldiers suspended it by a chain from the breast-plate. Some of these daggers had curious names such as 'sliced', 'eared', and 'kidneyed' because of the shapes of the handguards. The name 'left-hander' given to some daggers describes the hand they were held in. From about 1560 the *estocade-town buckler* was replaced by the small sword and dagger with cross-shaped guard. Its primary rôle was to parry the thrusts of the adversary's sword. It was particularly apt for fencing swordplay and swift counter-attacks. There are numerous variants of the dagger such as the *dagasse*, the *cinquedea* (so named because it was five fingers wide at its base) and the *daguette* (small dagger).

The work on the handles of knives, poniards and above all, the daggers, paralleled work on swords. Toward the end of the 16th century a ring was added to the cross-shaped guard to permit the holder to pass his index finger past the *quillions*. To this finger guard (also called pas-d'âne, literally donkey's foot for its resemblance to a hoofmark) was soon added a second ring for symmetry. The whole was called a *joining arch*. Later, a *knuckle bow* linking the *pommel* to the *quillions* was added to protect the hand from side blows. A ring perpendicular to the blade protected the forward part of the hand. Until the 16th century the guard became more and more complicated with added flourishes. Finally the idea came along of protecting the hand from points thrusts — a technique which had come to be used more fre-

Although these two French side-arms have only one edge, they are labelled as daggers. Both date from the second half of the 16th century. James de Rothschild Coll., Waddesdon Manor

This Swiss dagger dated 1567 comes with a small table service made up of a knife and skewer. The gilded brass scabbard is a forerunner of tableware sheaths. It is ornamented with a chased scene from the story of William Tell. Schweizerisches Nationalmuseum, Zürich

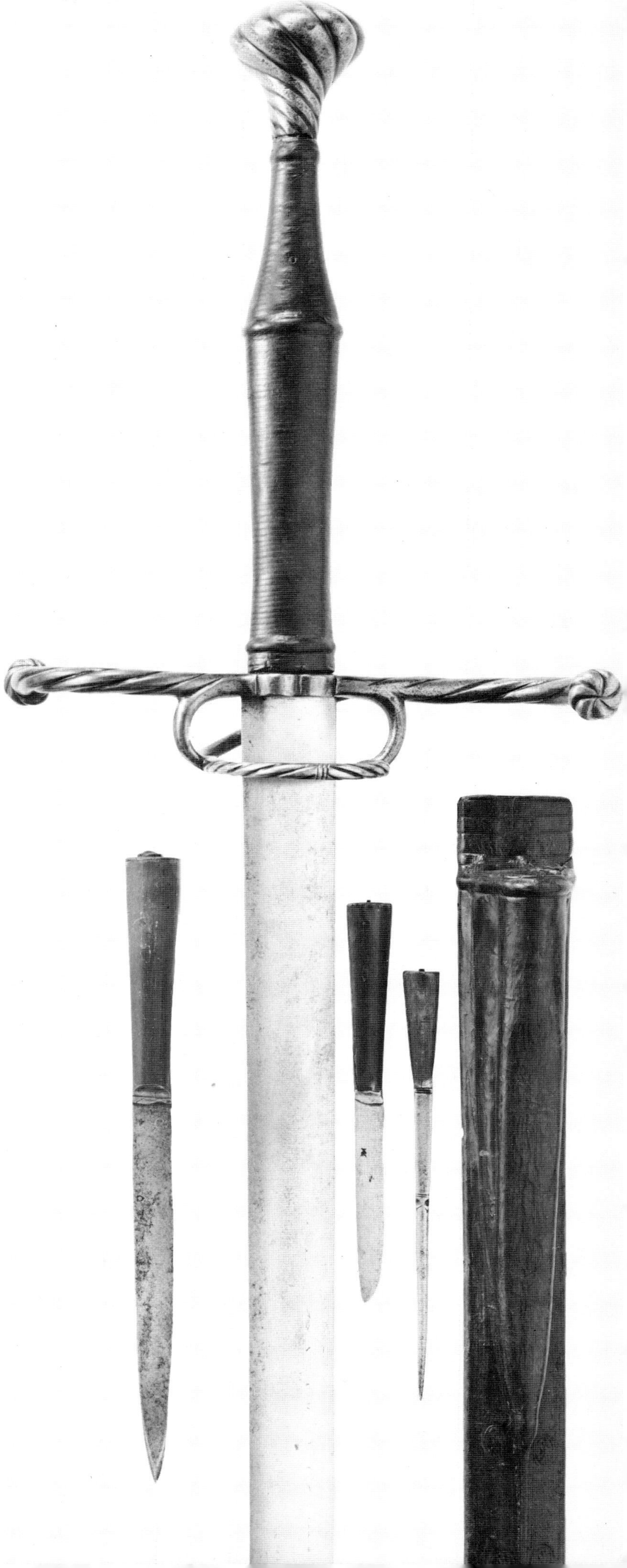

The scabbard of this sword, probably of Swiss manufacture, also contained two knives and a skewer with steel blades and bone handles. The set belonged to a Zurich citizen, Adam Näf, and dates from the beginning of the 16th century. Schweizerisches Nationalmuseum, Zurich

quently — and to do this a shell or cup guard was introduced. At first it was decoratively perforated forming a basket hilt, but it finished up as a solid piece which completely replaced the earlier protective devices.

Already in ancient times some knives, such as the *Mycenae dagger* with its gold incrustations, had become articles of jewelry. Closer to our times it is clear that the developments in handle making mentioned previously were motivated as much by aesthetic reasons as by practical ones. Scabbards, baldrics and blades grew together in richness of ornaments, adjunctions and inscriptions. Sometimes the craftsman would call in a jeweler for suggestions concerning filigrees, engraving and incrustations. In those days, the arms and accessories of war and of the hunt were considered precious objects and were thus highly decorated — sometimes to excess. Sometimes the motifs used were abstract, but they also included escutcheons, animal images, mottos and marks. The art of the great cutlery centers (Rives, Brescia, Toledo, Solingen, Sheffield) reached its height — in luxury as well as in technical perfection — in the 15th and 16th centuries. The guards and counter guards of German, Spanish or Italian daggers — made according to the buyer's orders — marked the summit of this art. In the time of Louis XIV, however, the decline had already begun. Starting in the 19th century, except for some pieces made in the Western United States, ornamentation disappeared. Fashion has since moved to the naked form and the time of great ornamentation is past, except, of course, for ceremonial weapons.

One of the most beautiful kinds of decoration is the Damascus type, which is used more frequently for swords than for daggers. Damascus blades combine an excellent edge with a remarkable suppleness. They are sonorous and of marvelous texture. But what attracts the eye are the black and white moiré designs. This manufacturing process was a mystery for westerners for a long time. Perret, the famous 18th century cutler, seems to have been the first to succeed in imitating the Damascus blades of Syria and Persia. In the

Two hunting knives. The French one on the left is chased and decorated with small precious stones; it dates from 1710. The belt and frog are decorated with animals in appliqué work. Schloss Nymphenburg, Munich. The saxony porcelain hilt of the knife on the right imitates a stag's hoof, and has a hunting scene painted on it. James de Rothschild Coll., Waddesdon Manor

The variety of different forms of daggers and small swords, from differing countries and periods, is large. The first three below are Spanish with characteristic cross piece and triangular knuckleguard (the end of the 16th century or beginning of the 17th century and 1680). The fourth and fifth are Italian (16th century). The sixth is of German origin (16th century) and its purpose is to break the weapon of the adversary, hence the sword's teeth.

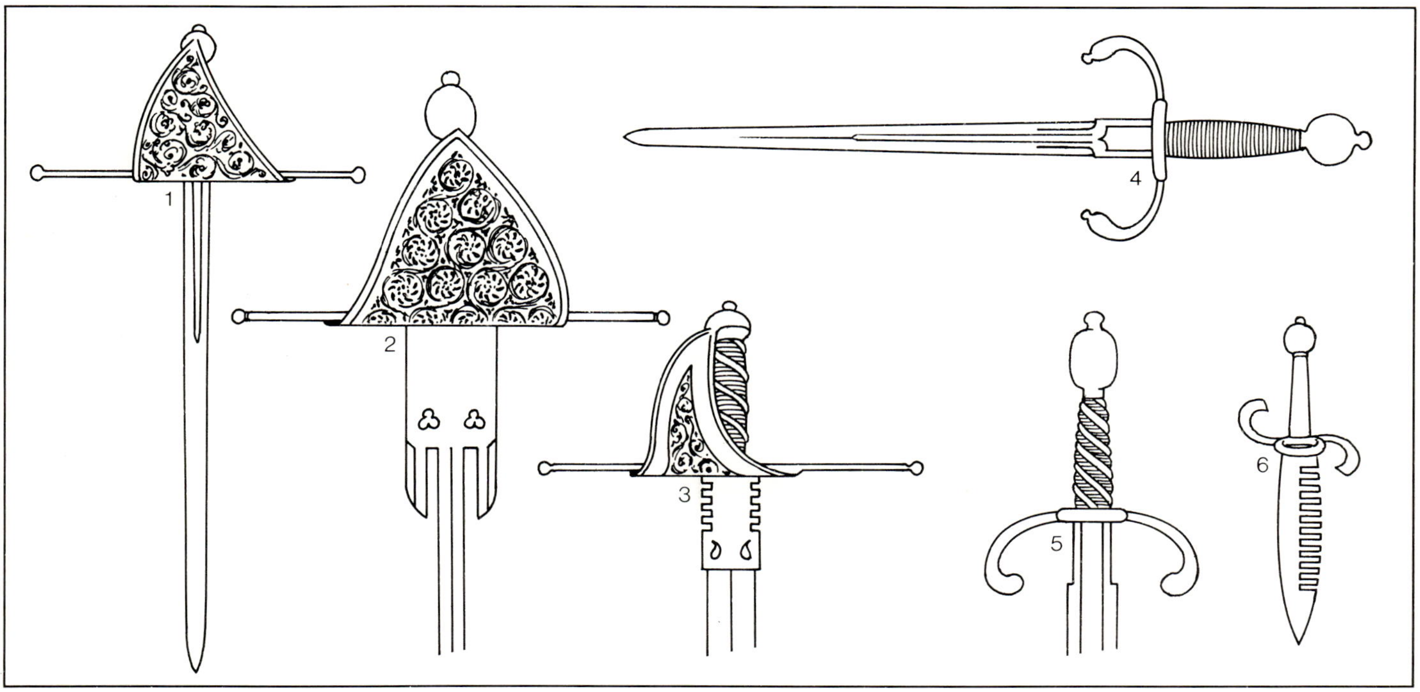

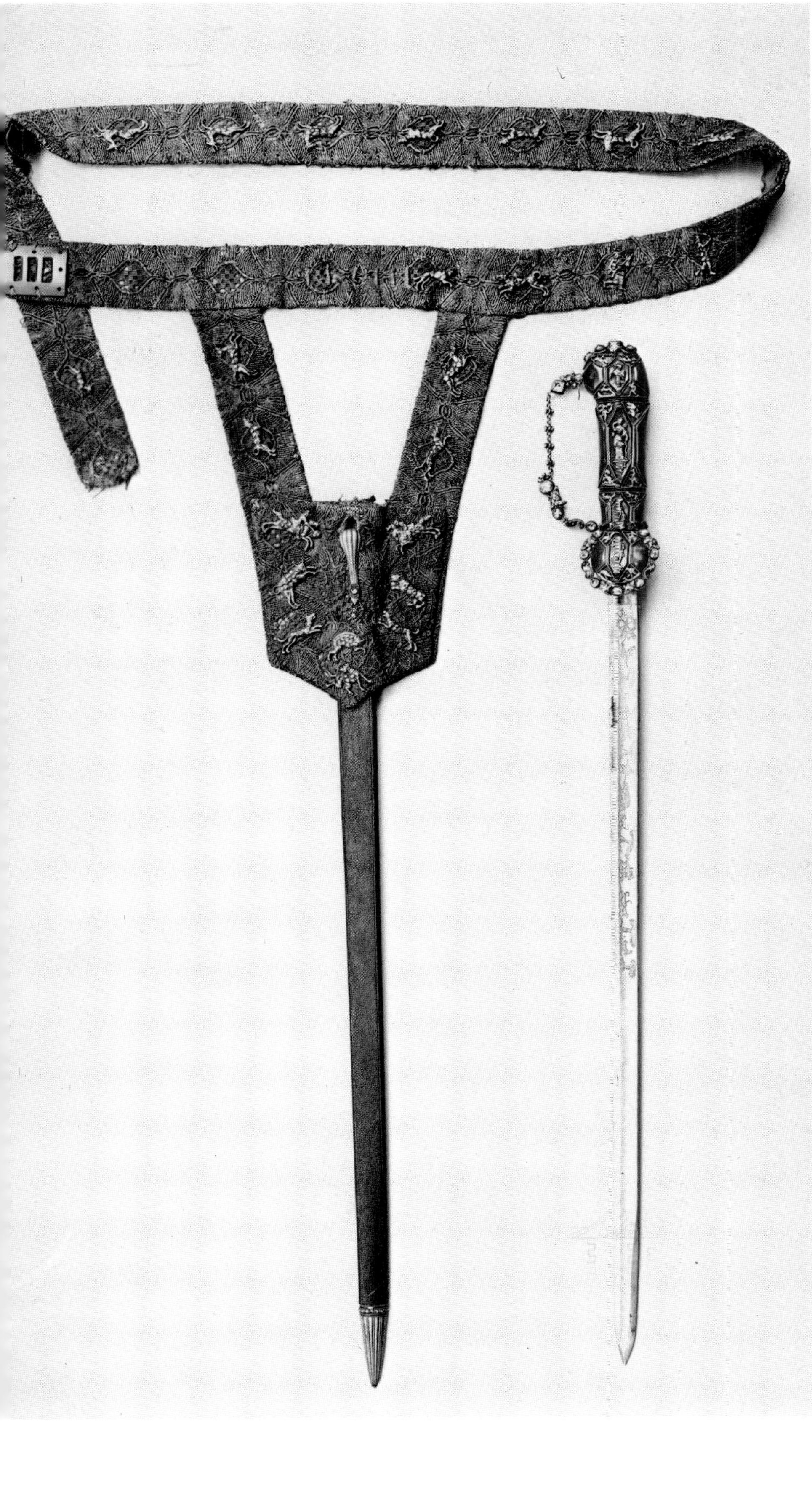

past, the Damascus surface was obtained by hammering heated 'spongy' iron without reaching the temperature of fusion. The result is a thin pancake that is free from dross, impurities and bubbles and whose surface is irregularly face-hardened from contact with the carbon of the heating matter. The hammering can also create veins of steel inside the metal. This thin slab is then etched with a graver in spirals or zigzags and flattened to become the rough shape of the blade. The two edges of this blade, the cutting edges, are case hardened and tempered. The center, which has become a mixture of different types of steel, is also thus different shades of grey and white, creating the much admired moiré patterns. In the etching process the concentric oval designs repeat themselves in widening circles. But there are many false Damascus objects; they can be made by drawing figures on the surface with a greasy material and then immersing the object into an acid which etches the areas not protected by the grease.

The cast bronze handle of this knife is a model of an Indian. The blade's section is diamond-shaped. George Wostenholme, Sheffield (early 20th century?) City Museum, Sheffield

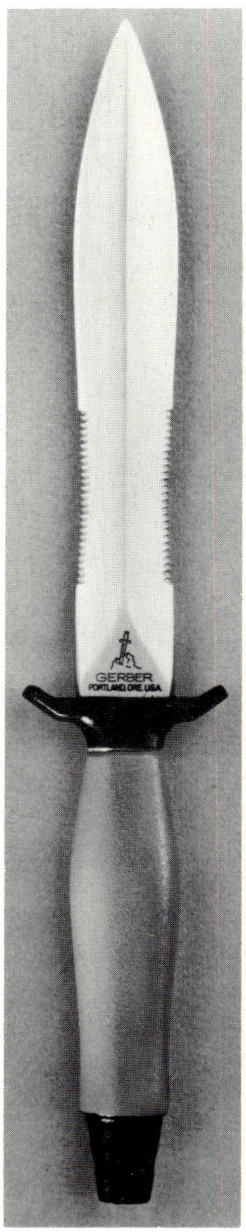

The Mark I on the right and the Mark II are U.S. Navy and Marines' survival knives. They are meant both for fighting and as general utility knives. The teeth of the Mark II are for cutting rope. The Mark I can be carried on a belt or attached to a boot. Length of the Mark I: 9 inches; the blade is 5 inches; length of the Mark II: 12 inches; the blade is 5 inches, thickness of both is 1/8 inch. Gerber, Portland, Oregon

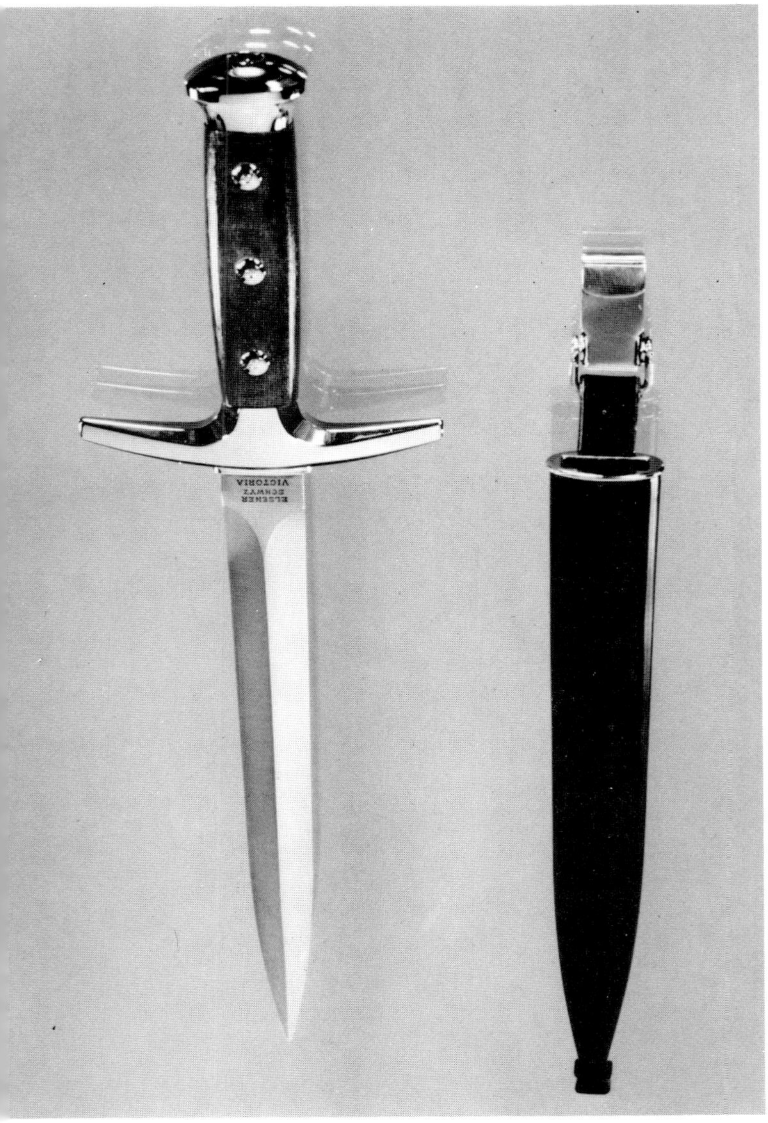

The Swiss officer's dirks are made in Schwyz, a Swiss county. The wooden handle is usually decorated with a tassel. The scabbard is steel. Coutellerie du Mont-Blanc Collection, Geneva

A knife of a Scottish Highlander anti-tank regiment fits into the soldier's stocking. The handle is wood, the pommel and the chapes of the sheath are metal. British manufacture. Coutellerie du Mont-Blanc Coll., Geneva

The knuckle-guard of this frogman's knife pivots and in so doing locks the knife to the metal scabbard. The rubber belt is used to attach the knife to the thigh of the swimmer. German made. Coutellerie du Mont-Blanc Collection, Geneva

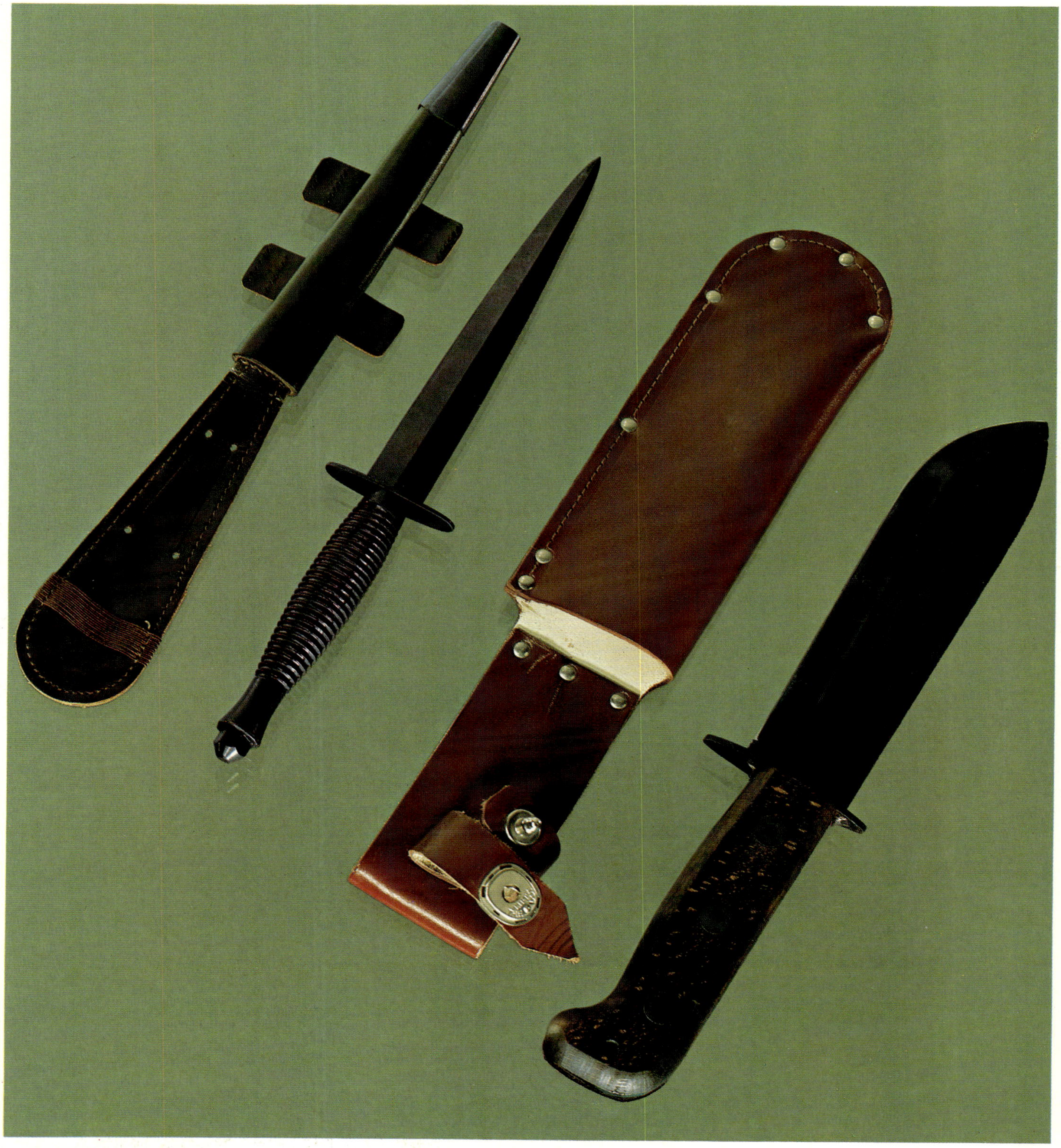

These two English knives were produced in the Second World War. The first is an all-steel commando dagger. The second is a survival type, for combat and use as a machete. This weapon was very useful in the Asian jungles. Both weapons are covered with a black protective varnish. Coutellerie du Mont Blanc Collection, Geneva

Right: The 'Randall Attack-Survival knife.' The handle is 1 inch in diameter with a waterproof cover which can contain medicines, bandages, etc. and which can transform the knife into a lance when fitted over a pole. The guard and cover are made of brass. The sheath contains a whetstone. The blade is almost 8 inches long. Another example of this 'model 18' is in the Museum of Modern Art, New York. Coutellerie du Mont Blanc Collection, Geneva

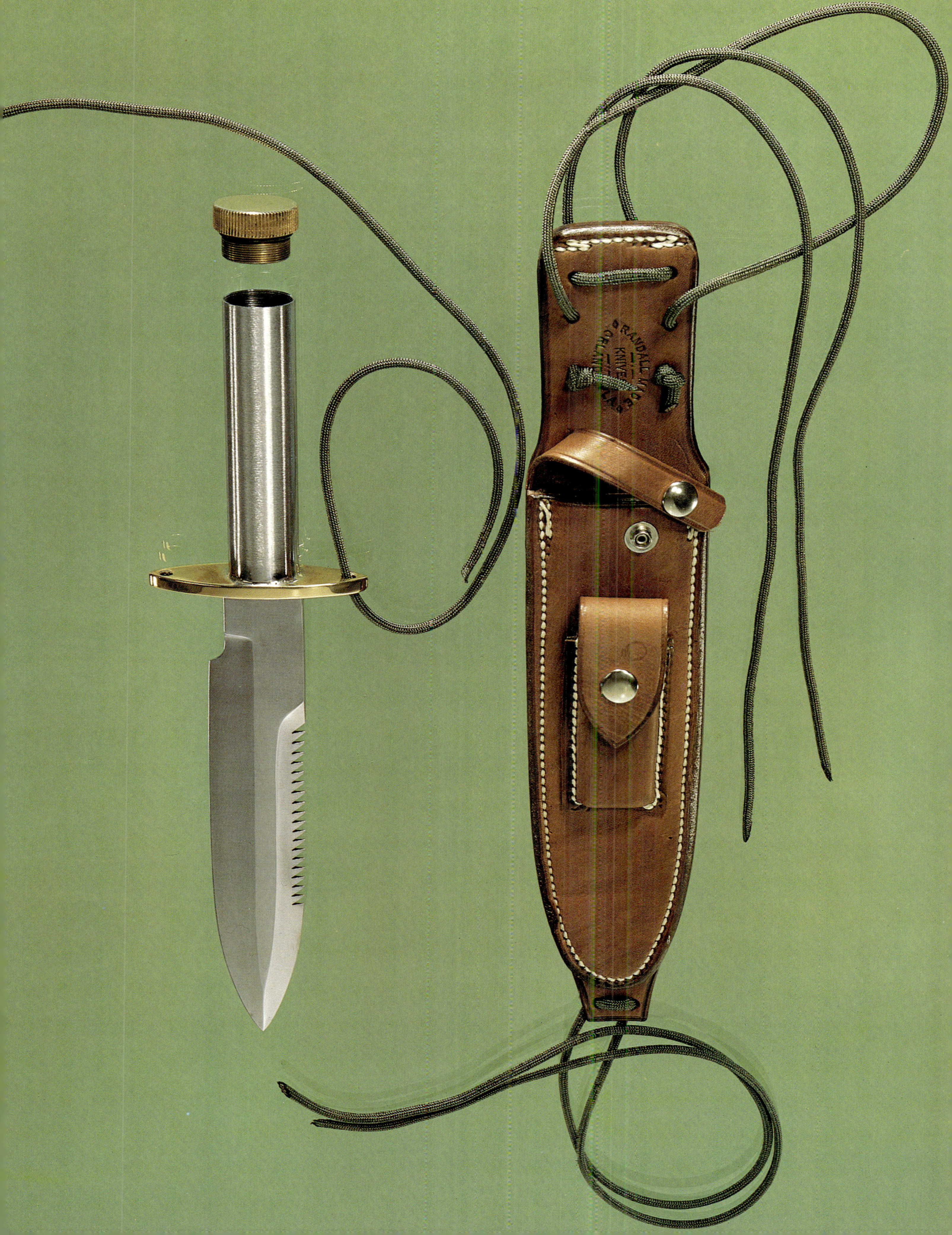
RANDALL

The Bowie and modern outdoor knives

Outdoor knives are enjoying a new wave of popularity. They can be found in all forms and for all purposes, the most beautiful often being handmade by craftsmen. The most famous among them is the *Bowie knife*, which has become a legend and has even been glorified in a film, *The Iron Mistress* (1951) which recalls the life of James Bowie. The *Bowie knife* is the ancestor of most of today's *hunting, fighting, diving*, and *scouting knives* with straight, wide blades whose edges curve towards the point from which a false edge extends backwards over a third of the spine.

During the conquest of the West, three weapons were born at about the same time: the Kentucky rifle, which, like the Pennsylvania rifle, is derived from the Tyrolian stutzer, but has an iron barrel instead of a steel one; the Colt revolver; and the *Bowie*. Not much is known about the inventor of this famous knife. Born in 1795 in Georgia or Kentucky, he made a fortune in the slave trade, smuggling, and perhaps piracy, before settling in Texas, which then belonged to Mexico. According to tradition, in 1830 or 1831, some five years before his death, he created the knife which bears his name. Bowie is said to have asked a highly skilful blacksmith named James Black, of Washington, Arkansas, to make him a knife according to a wooden model he had carved out of walnut. It is claimed that Black knew the secret of Damascus steel. However that may be, he is said to have made the knife in one night. (Today the renowned smithy has been rebuilt.) The knife was quickly attributed with magical powers. Shortly thereafter, Bowie, armed only with his knife, fought off three outlaws.

Prior to that time American knives had often been made out of scrap iron, old files and carriage springs, or used sabre-ends. America had not yet been able to produce anything in large quantities other than steel, but it certainly knew how to reuse the best European materials. The frontiersmen, be they pioneers, trappers, settlers, or outlaws, needed a solid, dependable *sheath knife* with which to cut game or bread, hack a path through the wilderness, or defend their scalps.

There are many kinds of knives but most can be reduced to several fundamental types. The whole series below is characterized by the fact that the center of gravity of the blade is situated towards the part of the spine that swells out, which gives the knife a particular advantage for cutting branches, for example, because it concentrates the striking force. The keen point will penetrate any hide; the guard protects the hand particularly well. The original 'Puma-Waldblatt' (1) has only a single edge. This knife, which measures 13 inches long (the blade is 9 inches) is ideal for dressing animal carcasses. In the 'Puma Rüdeman' (2) the flat of the blade, which can be seen towards the end, can be used as a hammer. The 12-inch knife (the blade is over 6 inches) is especially made for finishing off a wounded animal. The 'Puma White Hunter' (3) has a double edge — the left one can be used as a spare for those tasks that could be damaging to an edge. This model (length 11 inches with a 6-inch blade) also has a saw (on the left edge) and a notched section (on the left) which acts as a thumb piece. The next knife, the 'Puma-Cougar' (4) has many of the advantages of the 'White Hunter' and the blade also has a hooked portion at the end which is very useful when skinning large game. It can also be used to cut down reeds when duck hunting. The last knife (5) is distinct from the others and barely belongs with them. It is a fishing knife (length 12 inches with a 6-inch blade). Based upon the Puma catalogue, Solingen

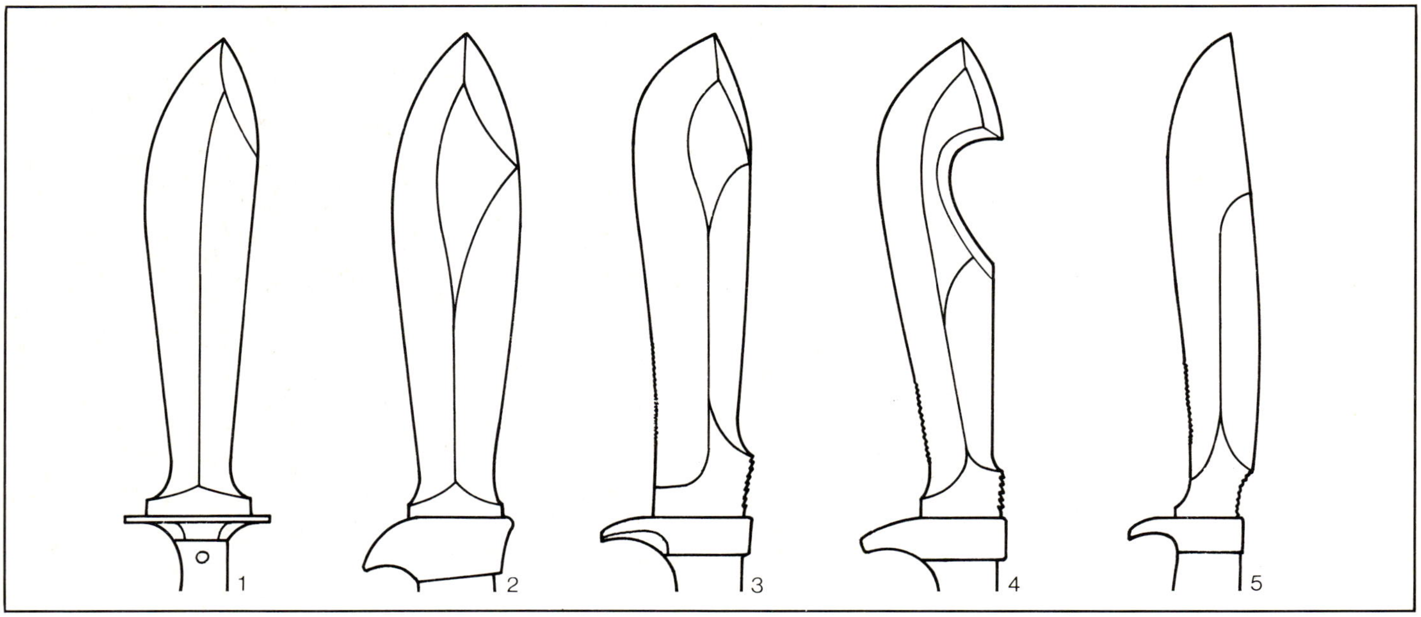

Rade Hawkins is the very prototype of the American craftsman-knifemaker: nature lover, hunter, fisherman, and sportsman. He always works by stock removal from blanks. The selection above includes relatively short-bladed outdoor knives as well as 'Bowie-type knives' and hatchets. Rade Hawkins, Red Oak, Georgia

Essentially Bowie improved the single-edged knife by combining the advantages of two types of arms. His knife had the profile and the balance of a fencing blade, which is generally light-weight, and the strength of a hunting knife, which is by definition a heavy weapon. No known knife of the 1830s or even 1840s is of certainty a *Bowie*. Thus we are forced to make conjectures. Estimates of the length of the *Bowie knife* vary between 12 and 18 inches. It must have been around two inches in width. The concave shape of the swedge would make it resemble a *yatagan*. These characteristics combined in one knife would make it suitable for thrusting, thanks to its strong and sharp point, but also for fearsome backward slashing. Davy Crockett is said to have declared upon seeing it: "It'll give you a stomach ache just to look at it." Bowie's fame could only grow. American cutlers executed a large number of orders for more or less faithful reproductions of the original. Soon the national industry could not handle the demand and foreign manufacturers from Sheffield, Solingen and Nogent were called in. Among the English *Bowies* is the I-XL of Wostenholm & Sons, Washington Works near Sheffield, of which thousands were exported in the last century, and which is still being made (a 10 inch blade, total length 15 inches, weight: 1 lb).

The many bloody incidents for which the knife was responsible, even in the halls of Congress, caused its prohibition, at least on paper, in a number of states — but this only increased its notoriety. Pioneers and outlaws were the prime customers for the *Bowies*. During the Mexican-American War the Yankee combatants, almost to a man, each had his own. The Confederates later also used it widely in two versions: the *Knuckle Bow*, so-called because the guard was joined to the pommel as in a sabre, and a *Bowie-bayonet*, with a ring. The use of the *Bowie* spread from the United States to Europe and thence to various parts of the world. It became the knife *par excellence* of the hunter, the explorer and the adventurer.

Although the conquest of the West and the development of the revolver brought about the decline of the *Bowie*, it has not yet disappeared. Imitations and copies are still being made.

The experts are far from agreement on the characteristics of the *Bowie knife*. The recognizable type is of a knife with a wide and heavy blade. The curve of the blade makes a point when it joins the swedge or the false edge. Some hold that the name *Bowie* — and a large number of knives claim this title — should only be given to those that have a point in the shape of a crescent. But most *Bowie knives* do not conform to this specification. Another controversial aspect is the existence, on the right side of the back or spine of certain combat knives, of a band of copper or brass which projects out in the shape of a dovetail. Its function is to impede the adversary's blade and keep it from sliding up the back of the knife, unpleasantly close to the hand of the fighter. There is no unanimity concerning the handle either: it can be made of a single piece mounted on the tang, or a set of laminated plates held together by rivets. It can be reinforced by a ring or bolster and have a metal pommel, but this is not always the case. However, everyone agrees upon the presence of a sturdy symmetrical guard with large *quillions* as well as a thick ricasso at the base of the blade. As it stands, the name 'Bowie' is applied today to a whole range of large single-edged knives with blades of from six to fourteen inches long.

Among the descendants and modern forms of the *Bowie* is the *Hospital Corps Knife*, which is sometimes considered a *bolo*. (There was a whole family of regulation US Army *bolos* in use up until the end of the Second World War. The *bolo* is a large, single-edged knife used in the Philippines. It is a kind of *machete* whose balance and design are in fact much closer to the Nepalese *Kukri* than to the *Bowie*.) As for the bayonet of the Krag rifle (1898), only its double usage distinguishes it from a true knife. Closer to our times, during the Second World War and the Vietnam War, some American troops were equipped with a knife derived from the *Bowie*, the *US M3*, which is a *trench knife*. The *Survival Bowie Knife* is made by the manufacturers Kinfolks, Collins, or Case. The US Navy and the Marines used, respectively, the *Mark 2* and the *K-Bar*.

The generic name of *fighting utility knives* confirms their double rôle of *combat* and 'all purpose' knives. For the last few decades the evolution of the general purpose *sheath knife*, to be used as much for scouting as skinning

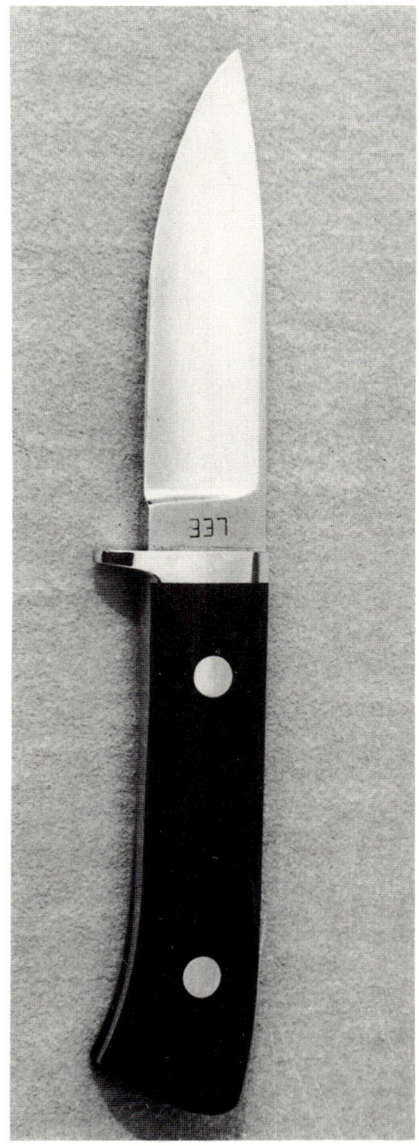
The handle of the 'Companion' by Lee, an outdoor and hunting knife, is made out of water buffalo horn. The blade measures 6 inches; the guard and rivets are of stainless steel. Tommy Lee, Gaffney, South Carolina

and dressing game, has brought about a reduction in the length of the blade and sometimes a different profile. But often the original *Bowie* design shows through. Sometimes the principal difference between the contemporary products and the traditional Bowie lies in the material of the handle — with synthetics replacing the wood, the ivory, or the stag-horn — and also in the quality of the steel.

Representative compositions of steels favored by American craftsmen are as follows:

		%			%
440-C	Iron	80.4	0-1	Iron	96.4
	Chromium	17.05		Manganese	1.35
	Manganese	0.5		Chromium	0.5
	Molybdenum	0.45		Tungsten	0.5
	Silicon	0.4		Silicon	0.35
	Nickel	0.2			

It can be seen that while the use of certain elements is constant, the proportion of chromium for example, can vary between 17 and 1.5 per cent. Some steels contain as many as nine components or more.

Another famous American arm is the *push dagger*, which appeared around 1848. As its name indicates, it is a fist-held dagger. It was the weapon *par excellence* of gamblers, and like the cards themselves, was easy to hide. Its large handle in the shape of a 'T' made it easy to hold, and the fact that only the short blade extended beyond the knuckles, gave it considerable force when thrust. A single stab would be enough to put the adversary out of the fight, often for ever.

The surgical steel (440-C) used in this skinning knife is particularly hard and corrosion-resistant. The handle is Macasar Ebony from Indonesia reinforced with steel and brass. Length 9 inches; blade 4 inches; thickness 3/16 inch. Gerber, Portland, Oregon

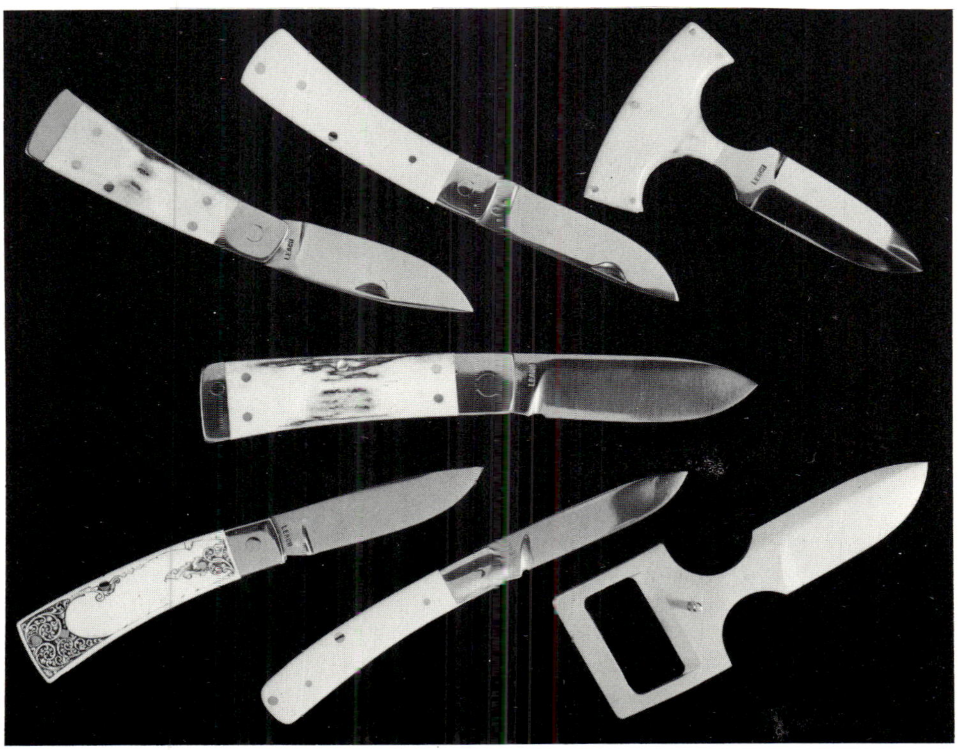

Leach's selection is remarkable for its variety — and this photograph only shows a few of the models! From top to bottom: a push dagger, five folding knives and one 'belt-buckle dagger'. All of the folding knives have one or more parts (such as the bolster and rivets) made of silver. The handle of the second is ivory. Milo J. Leach, Swartz Creek, Michigan

All of the camping sheath knives on the left, except the first, can have their handles tightened if the material of which they are made expands or contracts. The three center handles are made of horn while the first and the last are plastic. The sheaths are leather, except the first which is metal.

These three 'Bowies' are German. The blades are polished carbon steel, the handles are stag-horn. The point of the sheath is reinforced. Length of the blades: 6; 7; and 8 inches. Wingen, Solingen

In the 1920's, when most European cutlers were closing up shop or were about to do so, the United States enjoyed a renewal of interest in the craft. Among the pioneers of this new wave were Harry Morseth, Rudy Ruana and William Scagel.

The second step of this renaissance began in 1936 with W.D. Randall Junior. With the outbreak of war, Randall started forging knives for certain units of the U.S. Army. When peace returned, the same men who appreciated his arms under the colors began to place orders with him for hunting or outdoor knives. Today, the Randall team is made up of 16 people, but this does not eliminate waiting periods of many months for orders to be carried out because the knives are always custom forged. Nonetheless, this is far from the almost unbelievable waiting period of 10 years demanded by one cutler. The knife collector is therefore well advised to arm himself with patience, or to buy his specimens from a specialized dealer. Since the war, the number of full or part-time craftsmen has increased. The *Directory of Knifemakers* lists 110, but there are more. One of the most famous is R.W. Loveless, founder of the Knifemakers Guild and one of the first to use the 440-C steel. In Europe, there are no more than a handful of knifemakers of reputation in France and Germany.

The styles of the *outdoor knives* run to all shapes, from the stocky blades of some *hunting knives*, to the tapered blades of *fishing knives* and the hooked spine of *game skinning knives.* The most indefinite in shape are the general wilderness knives and those for general use when camping. Some of these are made with a bolt at the end so that the handle can be tightened if it should expand or contract under the influence of time and the elements.

Another very common type is the *throwing knife*, of which only the point is sharp. Contrary to common belief, *throwing knives* are not very practical for combat and rare is the thrower who can use them accurately. At a short distance it is possible to hit a target, but since the knife turns somersaults in flight, in order to hit the goal with the point it is necessary to know the exact distance and evaluate how many turns the knife will make.

The hatchet, together with the knife, is the most frequent companion of the camper. All of these pieces have stag-horn handles. Each of the two sheaths shown on the right carries a second smaller knife. Length of the blades: 8; 4; 5-1/8 and 2-1/2 inches. Wingen, Solingen

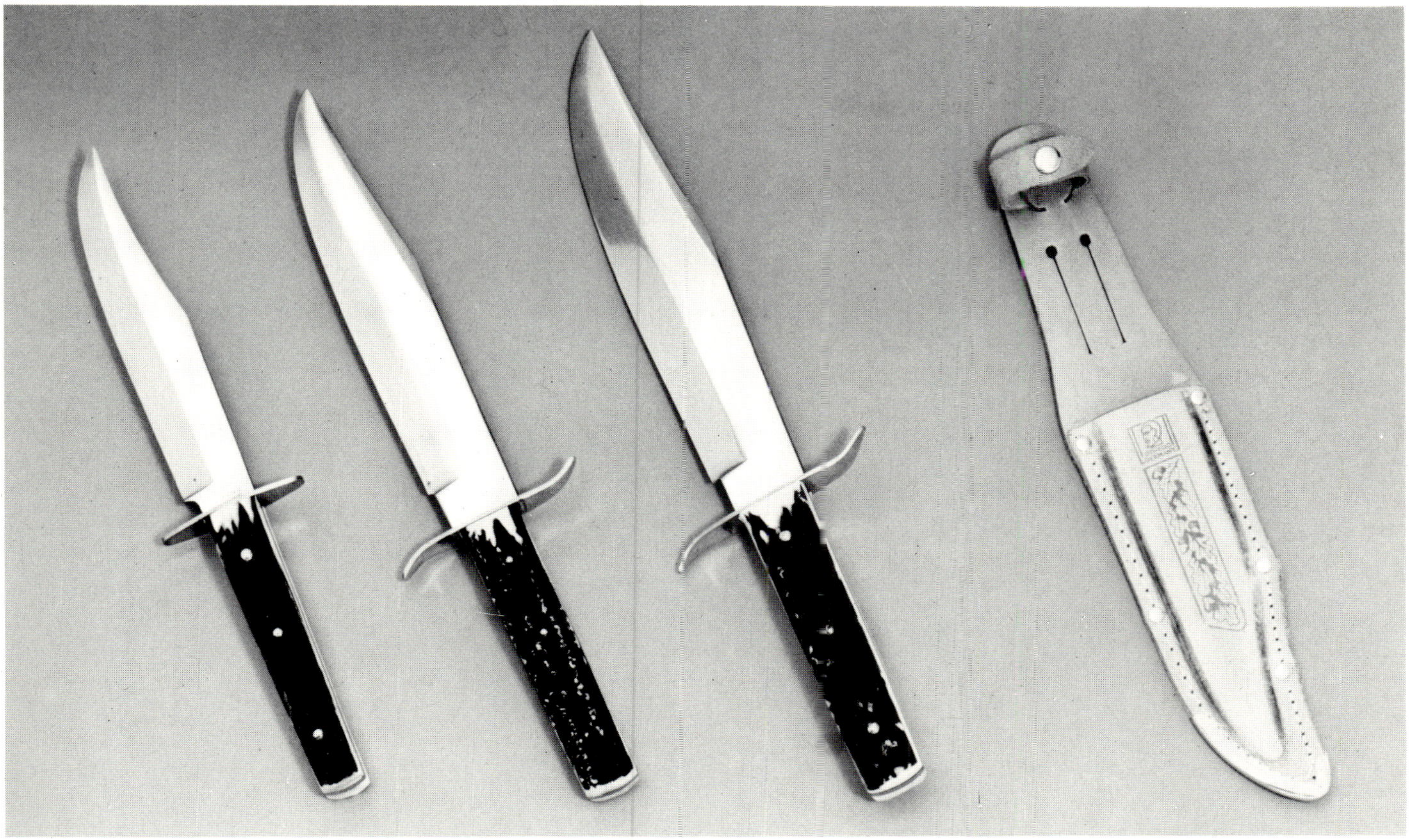

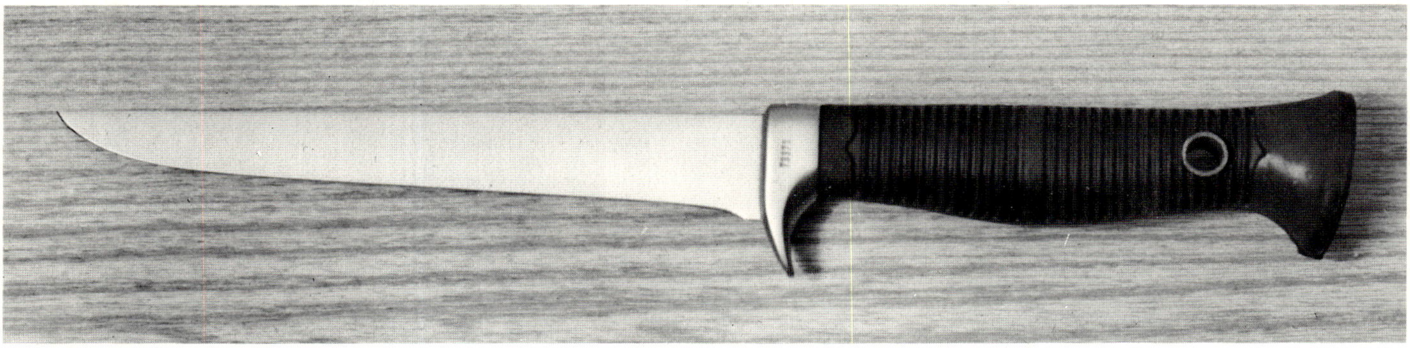

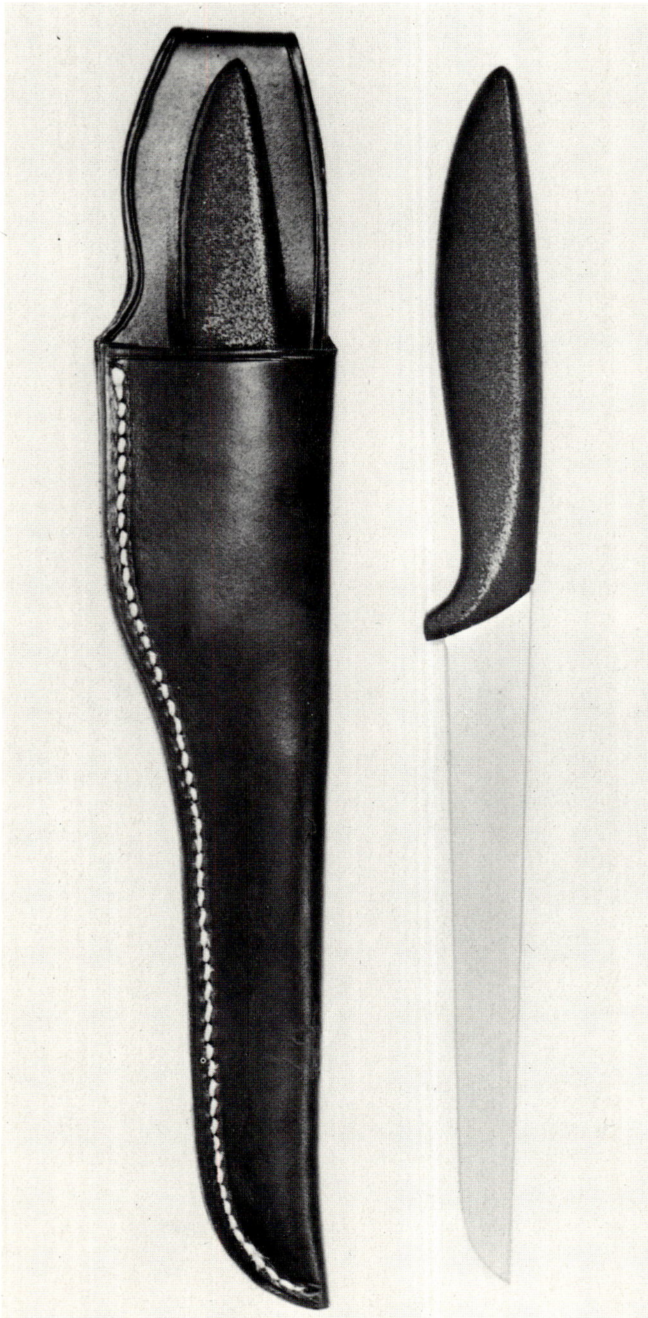

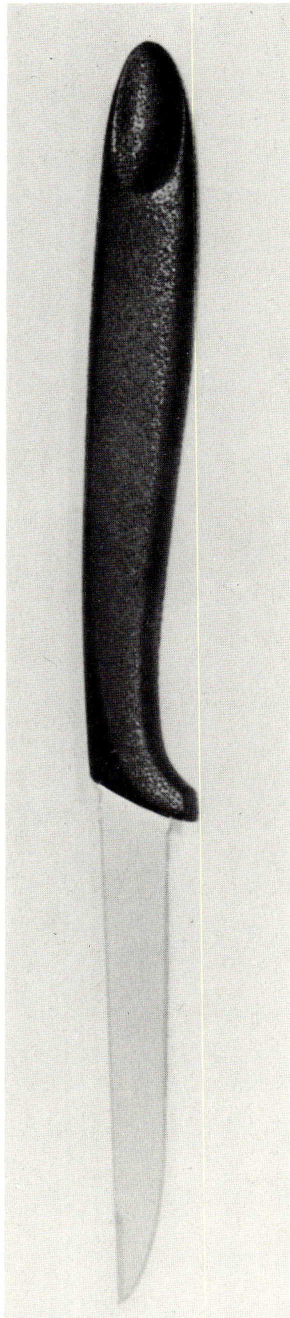

All of the knives pictured on this page are fisherman's knives, recognizable by the tapered blades. The 'Puma Mariner' knife (above) is especially made for filleting (length 11 inches; blade 6 inches). The blade is made of flexible steel, the non-slip handle is rubber. Puma, Solingen. On the left, the 'Muskie' (11-1/4 inches — 6-inch blade) is used for boning and filleting. The blade is surgical steel and both handles are of rough surfaced aluminium. The 'Trout and Bird' next to it is as useful to the trout fisherman as to the shooter of game birds. Length 8 inches; blade 3-1/4 inches. The hollow on the handle is particularly useful for gutting the fish.

The knife on top is imposingly called the 'Guthook Kangaroo Folder'. Only one of the blades folds into the handle, which is made of Micarta. Below it on the left is a hunting knife, the 'Buckle Blade Hunter'; the handle is plastic. The other two are push daggers with horn handles. The one on the right can be used as a bolo tie. Nolen Knives, Mercedes/Corpus Christi, Texas.

NK
THE RIGHT

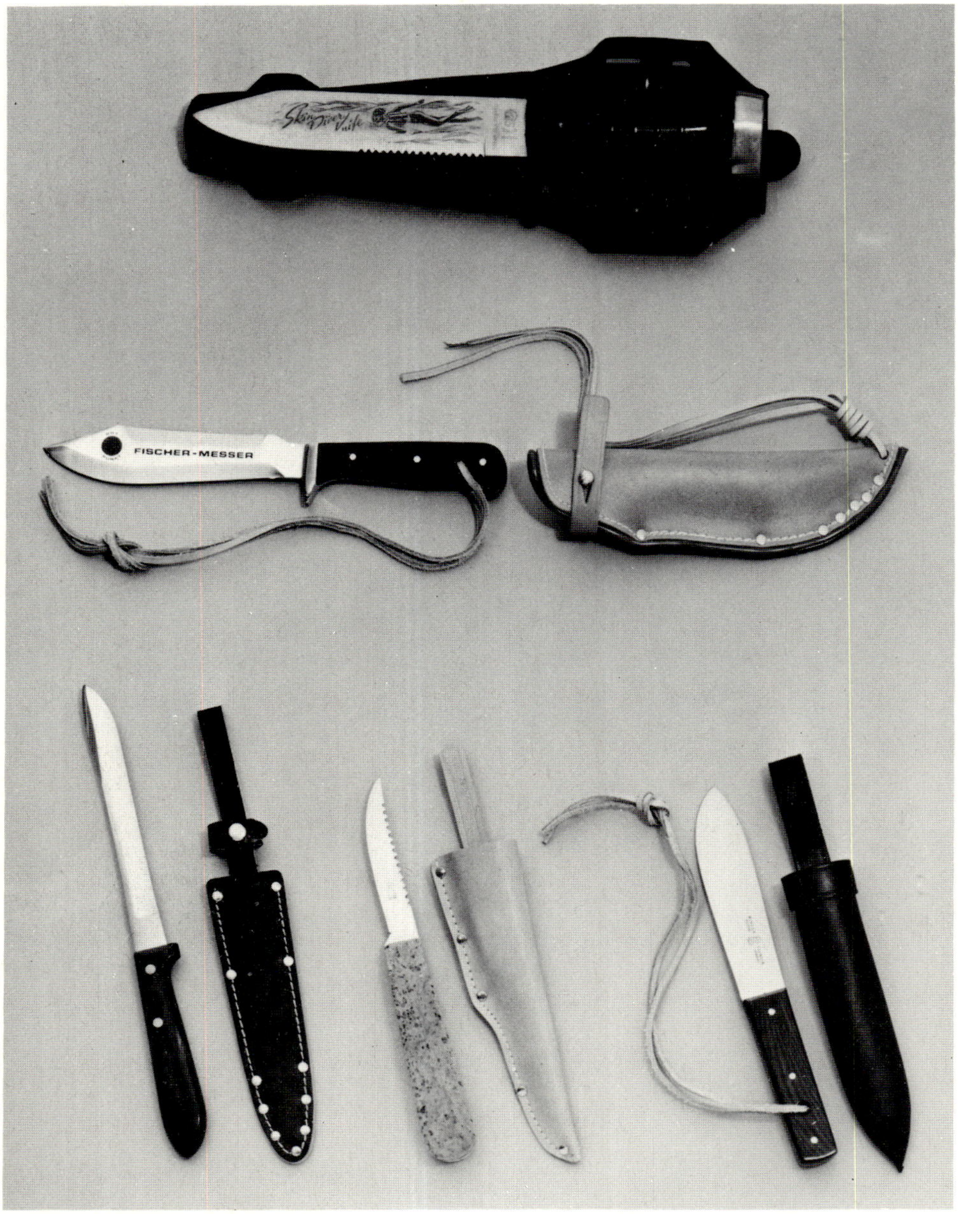

The first of these knives for sailors and fishermen have specially adapted handles. The first is non-slip rubber, the second and third are humidity-proof plastified wood, and the fourth is cork to make the knife float. The first is a diver's knife, the second a fisherman's, and the last three are sailor's knives. Wingen, Solingen

Right: The shape of the first knife, the 'Puma Hirschfanger' is adapted for bleeding game (length 15 inches — 9-1/2-inch blade). The guard is particularly classic. Like the others, the handle is stag-horn. The next two are an original 'Puma-Waldblatt' and a 'Rüdemann'. The diagram on page 38 can be helpful here. The little knife, the 'Puma-Nicker', can be used for finishing off the animal as well as for odd jobs. (Length: 8 inches — 4-inch blade) Puma, Solingen

Right: Most of the handles of these hunting knives seem heavy and out of proportion to the blades. But they provide a good grip and the impact of a blow is reinforced; these knives are more powerful than precise. The handles are stag-horn and some are partly gilded. Some have various accessory blades (such as the gutting blade in the last knife). Blade lengths: 3-3/4; 4-3/8; 4-3/8; 4; 4-3/8 inches. Wingen, Solingen

From top to bottom: a boot-knife or dagger which can be concealed in a boot; for this reason the handle is extremely flat and the thickness of the blade is not more than 3/16-inch. Facing it is a featherweight skinning knife (weight: approx. 2-1/2 oz.) The knife in the center is also ultra-light; the handle is stag-horn (width of the blade 1-7/8 inches). The last, recognizable by its tapered blade, is a filleting knife. The blade is very thin, only 1/8 inch. John T. Smith, Southaven, Mississippi

LONS-MESSER

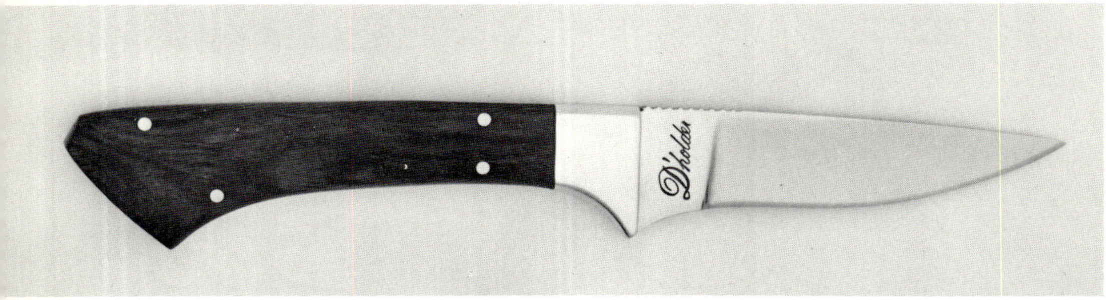

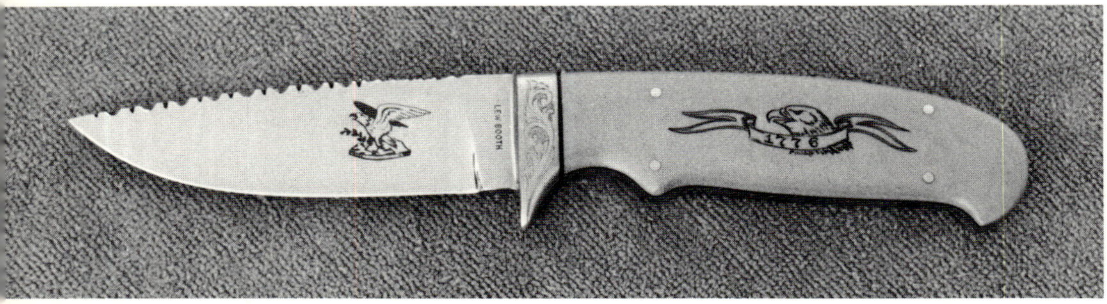

Top: This D'Holder knife is nick-named 'Kit' because with its small size (length of the blade 2-1/2 inches), it can be carried anywhere. It is a trout fisherman's or game bird hunter's knife. D'Alton Holder, Phoenix, Arizona

Above: This knife was decorated by the same man who forged it. Lew Booth keeps his manufacturing processes a secret — especially how he tempers the steel. Lew Booth, Boonton, New Jersey

Right: The form of this short (4 inches) blade knife is admirable, being at the same time functional and aesthetic. The rosewood handle was made to measure to the fingers of its owner. Corbet R. Sigman, Liberty, West Virginia

Below: Note the finesse of the blade of the top hunting knife (width 7/8 inch) and its shortness in relation to the handle. The handle of the second one is horn. Length of the blade: 4 inches; thickness: 1/4 inch. The handles of the first and third knives are plastic imitation wood. Dimensions of third knife same as the second. All three are full tapered tang models. Dwight L. Towell, Midvale, Idaho

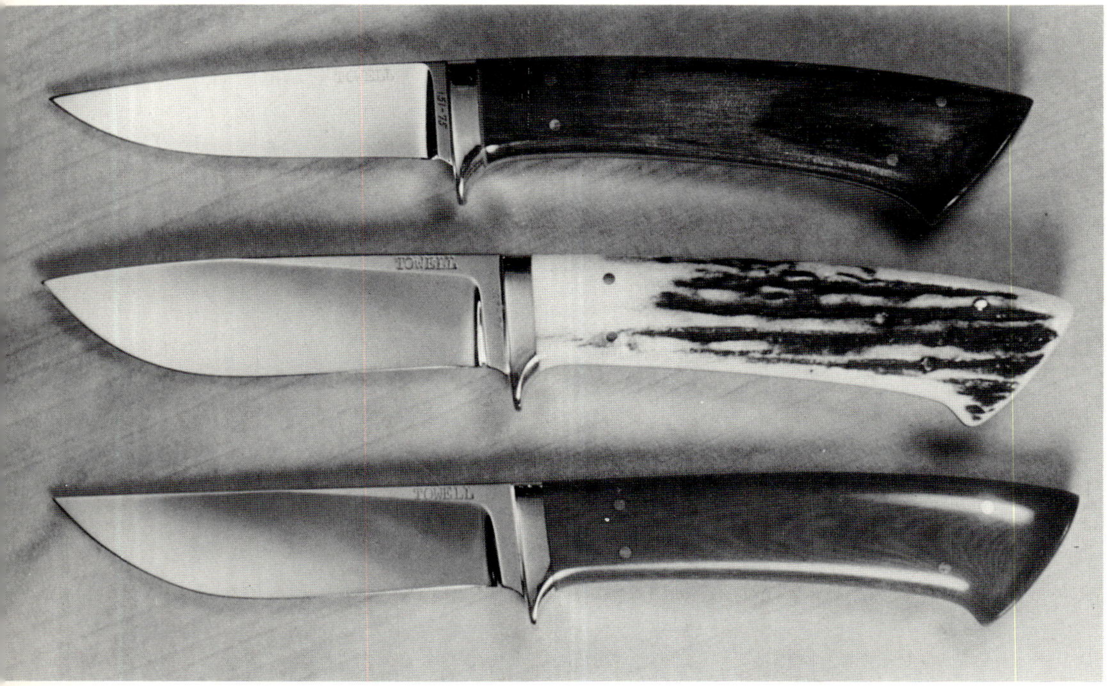

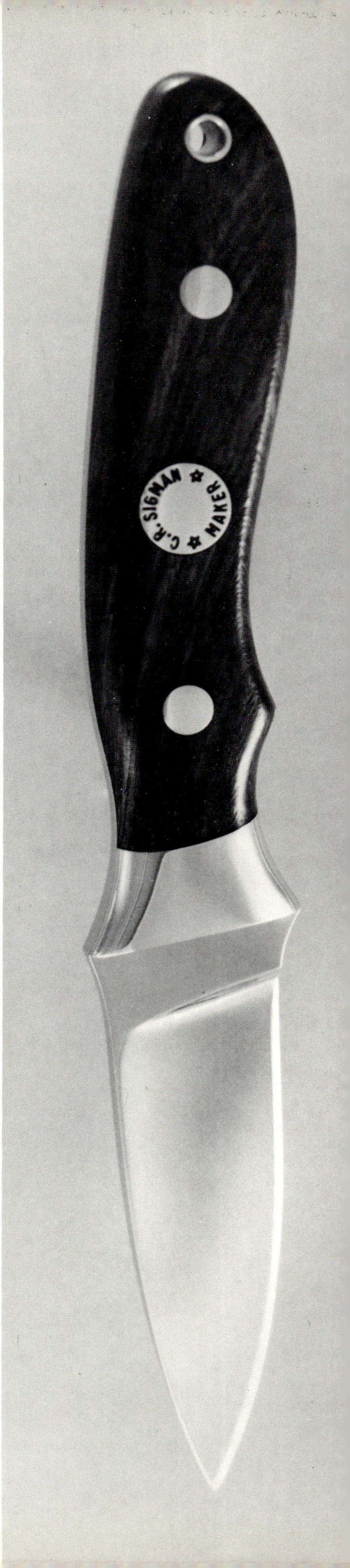

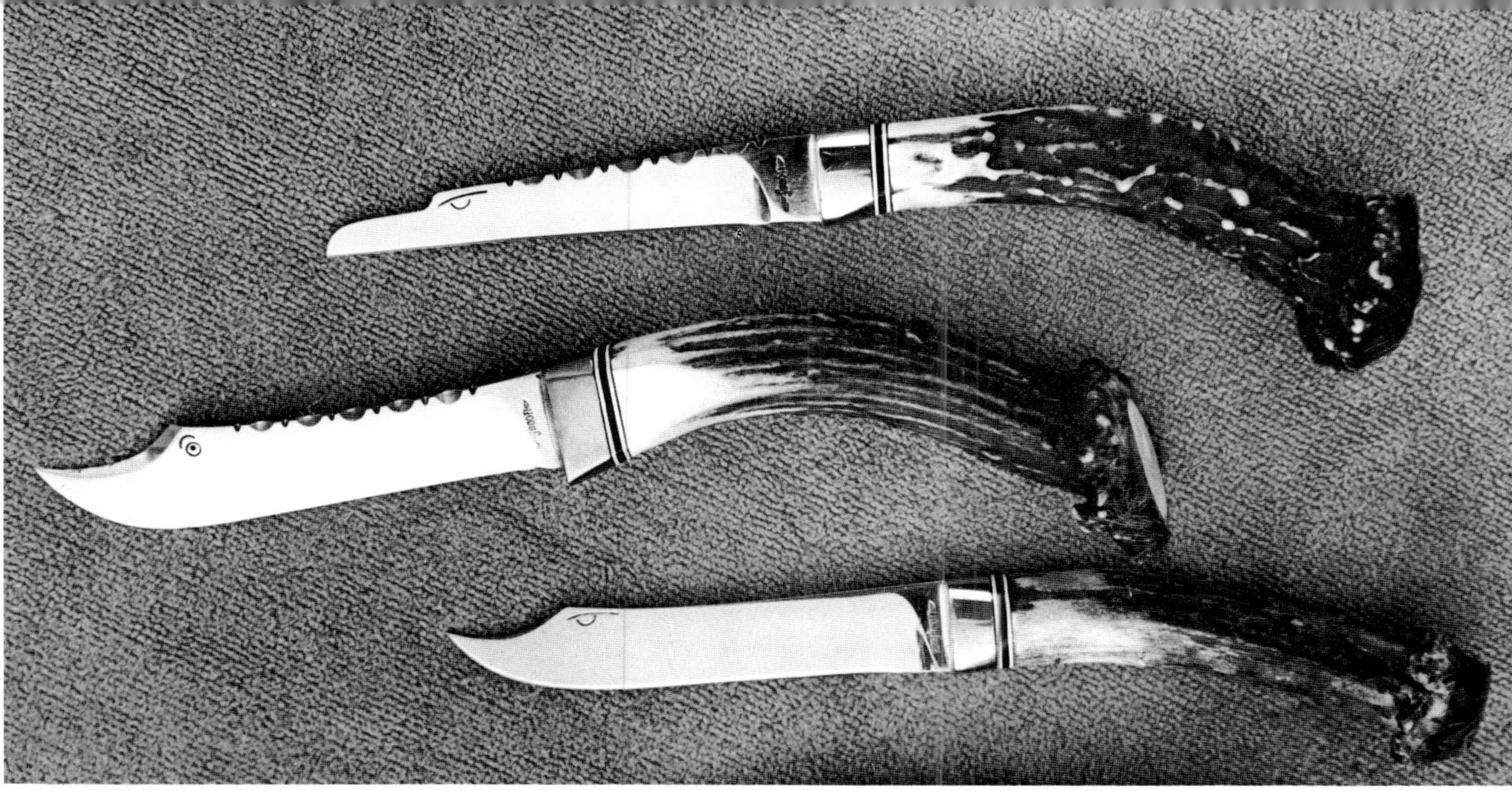

Some of Lew Booth's knives are very unusual in their use of the curved part of a stag-horn as handles, and the shape and decoration of the blades. Lew Booth, Boonton, New Jersey

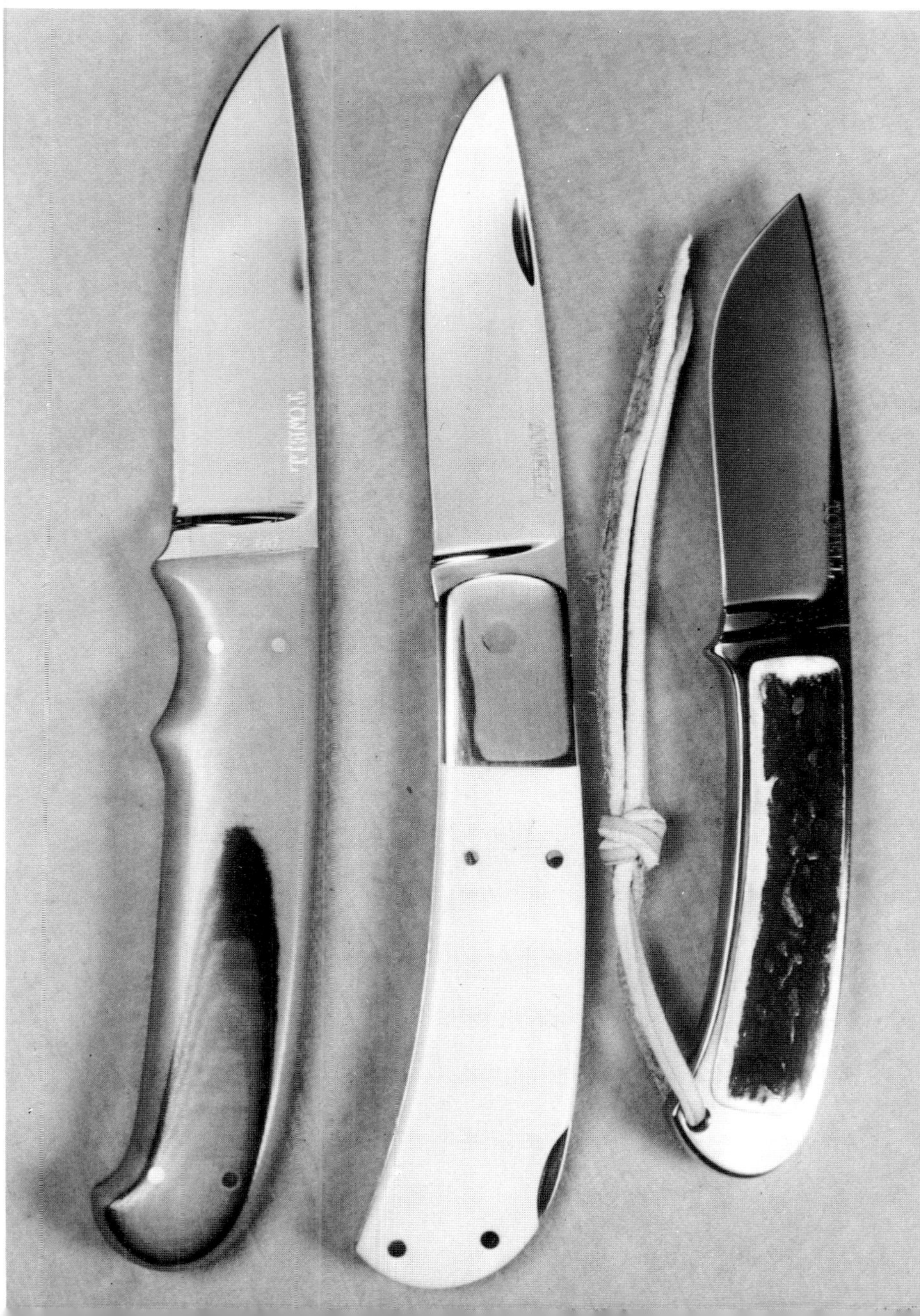

Right: Some American knifemakers make folding knives as well as sheath knives. The folding knife in the center has a safety catch, the thickness of the tang-handle of the third knife is twice the thickness of the blade (handle 1/8 inch — blade 2/8 inch). Length of the blades: 2-1/8 inches; 3-3/8 inches, 3-1/2 inches. Dwight L. Towell, Midvale, Idaho

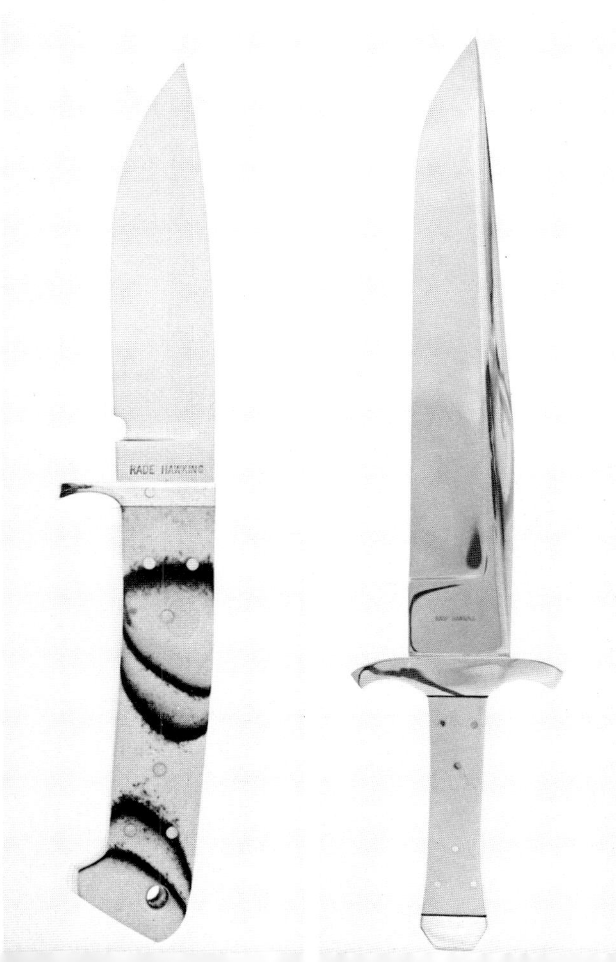

Two knives by Hawkins. The back of the first is a 'clip point' style which is often seen; the moiré handle is plastic. The second is a Bowie knife by the widest definition. Length of the blades: 4 and 12 inches. R. Hawkins, Red Oak, Georgia

Above, left: These hunting knives are called 'stiletts' in Germany. The stag hoof handle gives the one on the right a very picturesque appearance. Length of the blades: 4 inches. Wingen, Solingen

Above: This skinner by Colt has an impressive blade with a mirror finish. The disproportion between the length of the wood conglomerate handle and the blade is very marked. The guard is nickel-silver. American made. Coutellerie du Mont-Blanc Collection, Geneva

Left, from top to bottom: A 'Canadian Trapper' knife, two 'Bowie' style knives, two trapper's knives with raised points and two models for skinning buffalo. Some of the handles are made of exotic woods, and the others are stag-horn. Wingen, Solingen

Right: All of the folding knives with safety catch and fixed blade models grouped together here belong to the same style. A connoisseur can spot the family resemblance between the two types. The blades were polished on an 18-inch water-cooled grinding wheel after the tempering. The handles are made of cocobolo, Micarta, and Indian stag-horn. Centofante Knives, Tampa, Florida

CENTOFANTE
020
CENTOFANTE
001
CENTOFANTE
073

CENTOFANTE
CENTOFANTE

These two knives, one with a tapered and the other with a wide blade, are considered complementary and often carried in the same sheath. The handle is brown Micarta. Length of the blades: 2-1/8; 2-5/8 inches. Dwight L. Towell, Midvale, Idaho

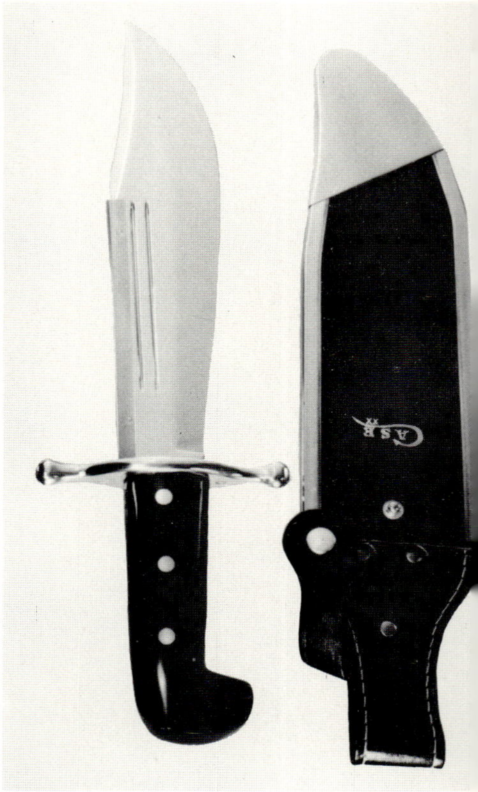

The spine of this Bowie Hunter by Case is covered by a piece of brass in the classic style of the first Bowies. The guard, rivets and chapes of the sheath are also of the same metal. The handle is synthetic. American made. Coutellerie du Mont-Blanc Collection, Geneva

The originality of these knives lies in their handles, which are made out of the jaw bones of brown and grizzly bears. Note the scalloped edges. Lew Booth, Boonton, New Jersey

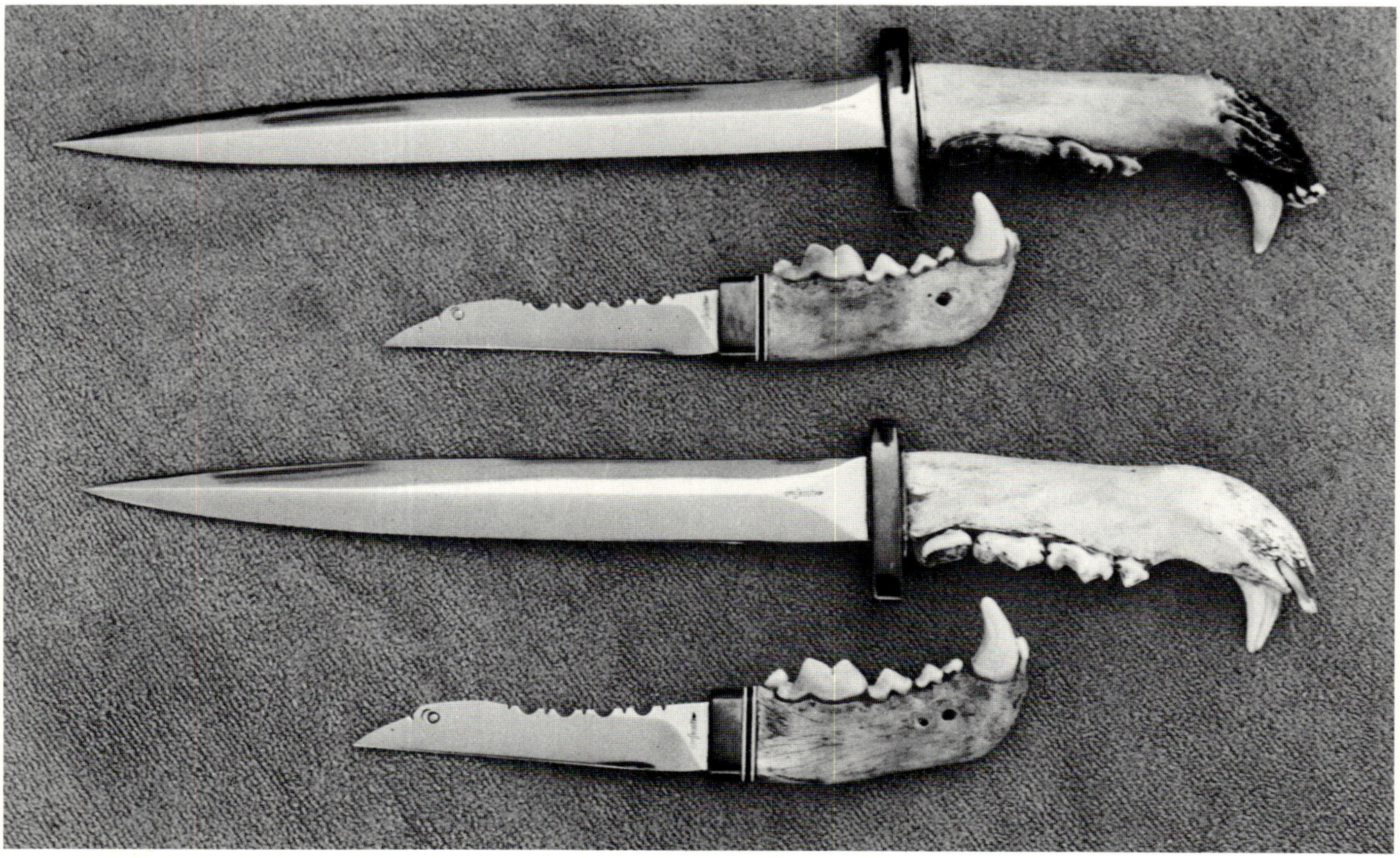

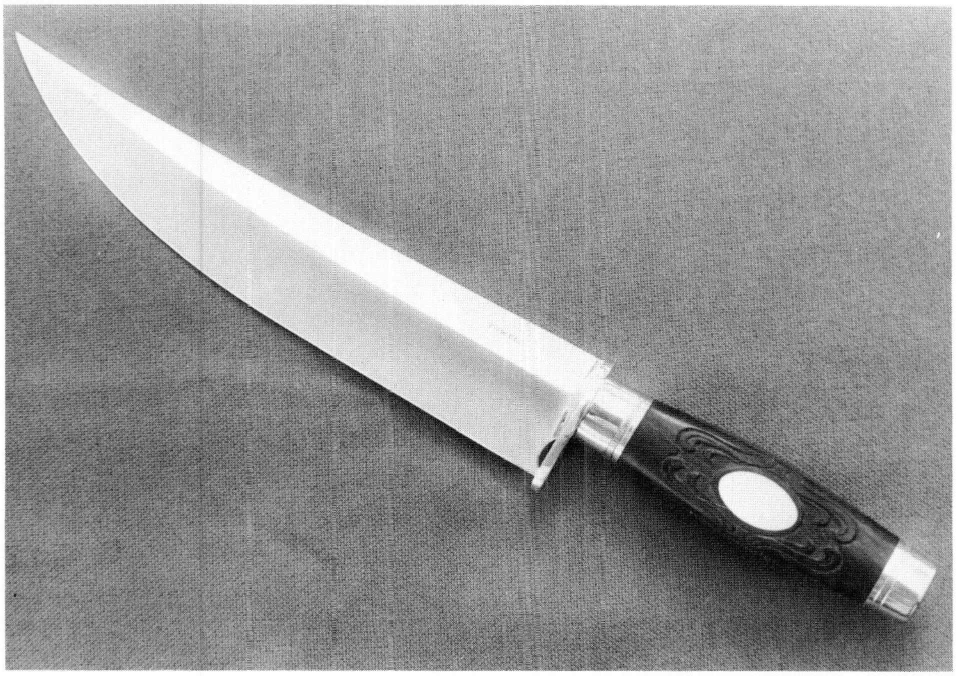

The story of James Bowie is mingled with legend. The fabled smithy of James Black, maker of the first Bowie, has been rebuilt in Washington, Arkansas. The shape of the Bowie has been the subject of innumerable discussions. Whatever it was, it was inspired by knives imported from Sheffield, the most famous brand being 'Green River'. To speak of an English Bowie as Holder does (on the left) is not a contradiction in terms. Moreover, nothing is missing, not the famous crescent shape of the point, nor the straight handle. The handle is of stag-horn. D'Alton Holder, Phoenix, Arizona. The model below is also a 'Bowie', but called by a different name. The handle is rosewood and nickel-silver. The model above, by the same craftsman, is called a 'Resin Bowie' which has nothing to do with what the handle is made of (which is carved Indian rosewood) but refers to the name of the older brother of James Bowie, often spelled 'Rezin'. Blade lengths: 10; 9-1/4; 8 inches. Dwight L. Towell, Midvale, Idaho

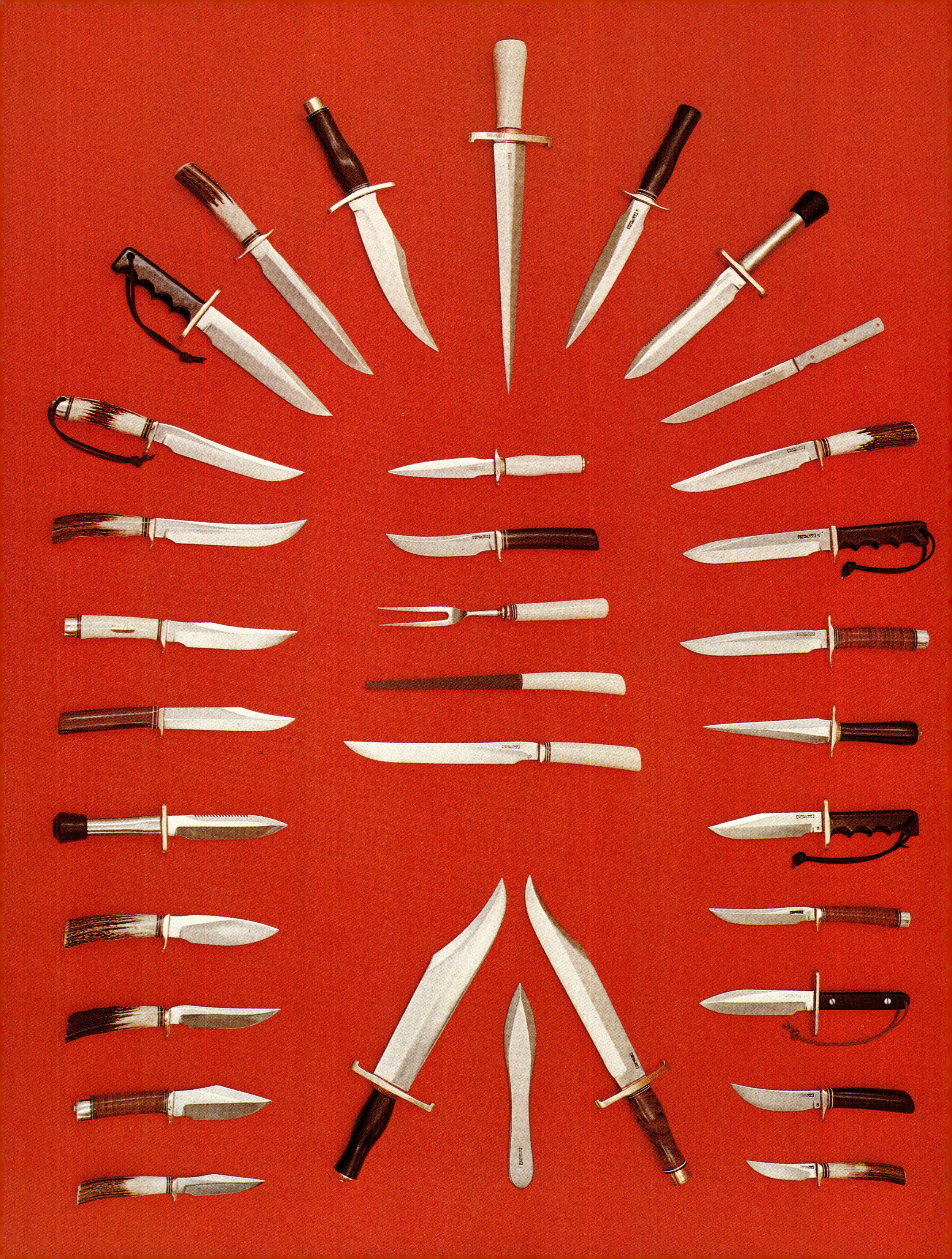

The handle of this knife is made of brown boco wood, which is so hard that it is sometimes classified as an ironwood. Corbet Sigman, Liberty, West Virginia

A 'Nolen Hunter' (a hunting knife with sambar wood handle), a 'guthook', a 'Rio Grande Skinner' with wooden handle, and a richly decorated 'Fancy Folder'. The Nolen knife is highly polished, but without a mirror finish. Nolen Knives, Mercedes/Corpus Christi, Texas

In 1936 W.D. Randall Junior began his career in knifemaking, and this date is often considered as marking the rebirth of knife craftsmanship, even though Randall was not the first to forge knives in the 20th century. Randall's knives were used by American soldiers in the Second World War and in Korea, but the blades also became famous in such peaceful pursuits as hunting, sports, and exploration, from the Arctic Ocean to tropical forests. Naturally, it was Randall that NASA asked to create knives for project Mercury. Randall Knives, Orlando, Florida

The elegant handle of this knife is Micarta, imitating bone. Note the large bolster. Corbet R. Sigman, Liberty, West Virginia

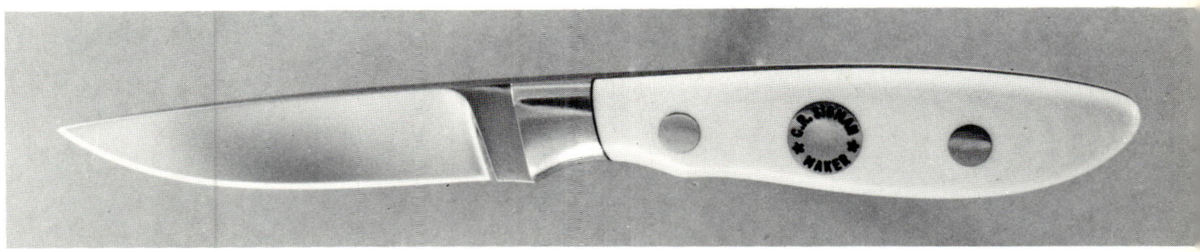

Holder has named this 'My Knife' which shows how much he thinks of it. It is a thing of great beauty: the handle is alternately African amber and fossilized walrus ivory with spacers of various materials, notably brass. Length of the blade: 4-1/4 inches; thickness of the spine: 3/16 inch. D'Alton Holder, Phoenix, Arizona

Right and below right: There is a family resemblance among these knives, and yet each one has a slightly different blade ranging from convex to concave. They are the 'Custom Cougar' with its leather sheath, the 'Cougar' with stag-horn handle; the 'Custom Eagle', with rosewood handle, and the 'Eagle'. The shape of the blade and its length are constant: 5 inches. Zack Knives, Pahokee, Florida

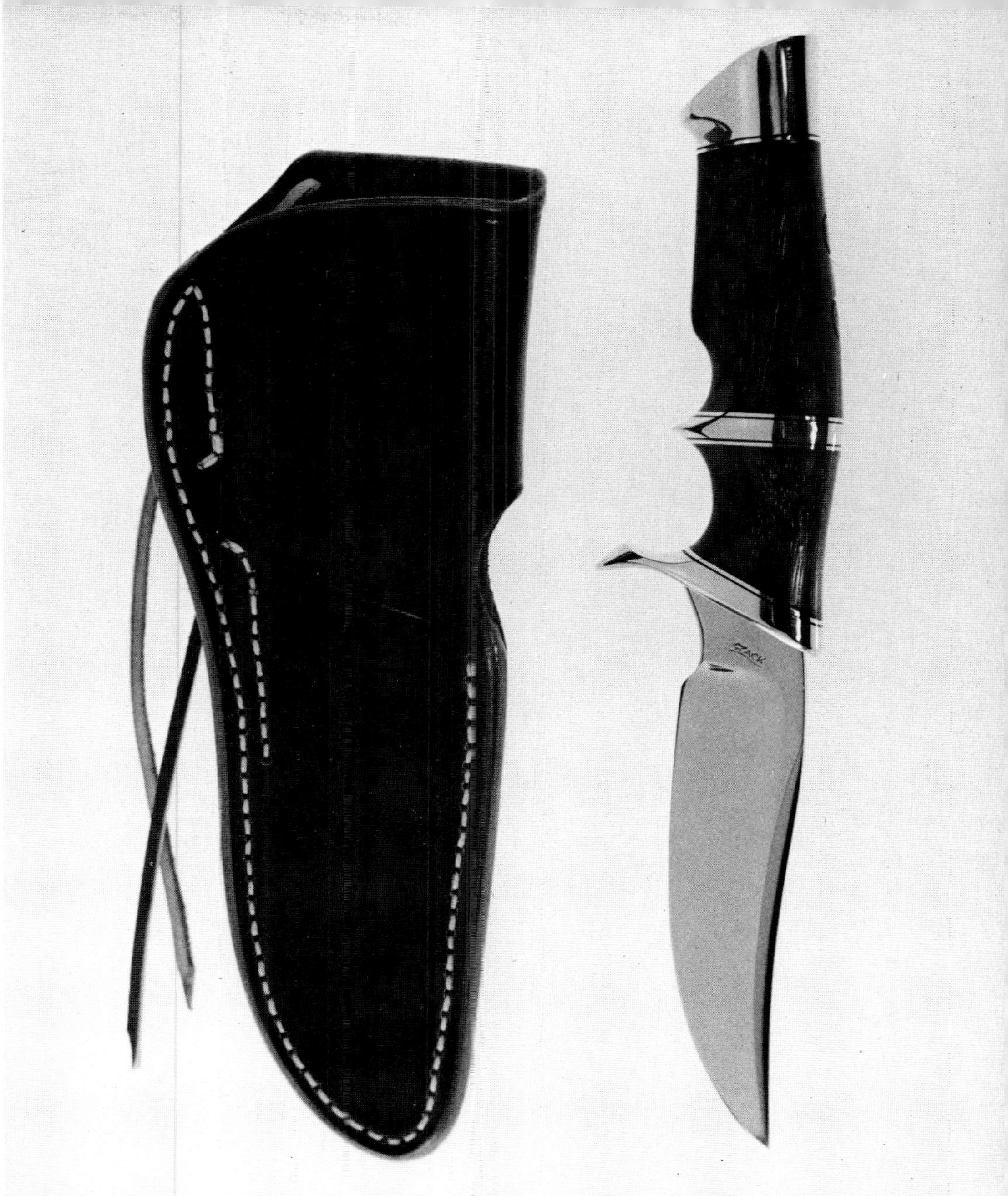

Lew Booth, who designed both of these knives, has received many prizes at exhibitions. He makes one or two knives a week. For his handles he uses bone (on the left), stag-horn (on the right), ivory and grizzly jaws. He engraved these knives himself. Note the peculiar shape of the one on the left. Lew Booth, Boonton, New Jersey

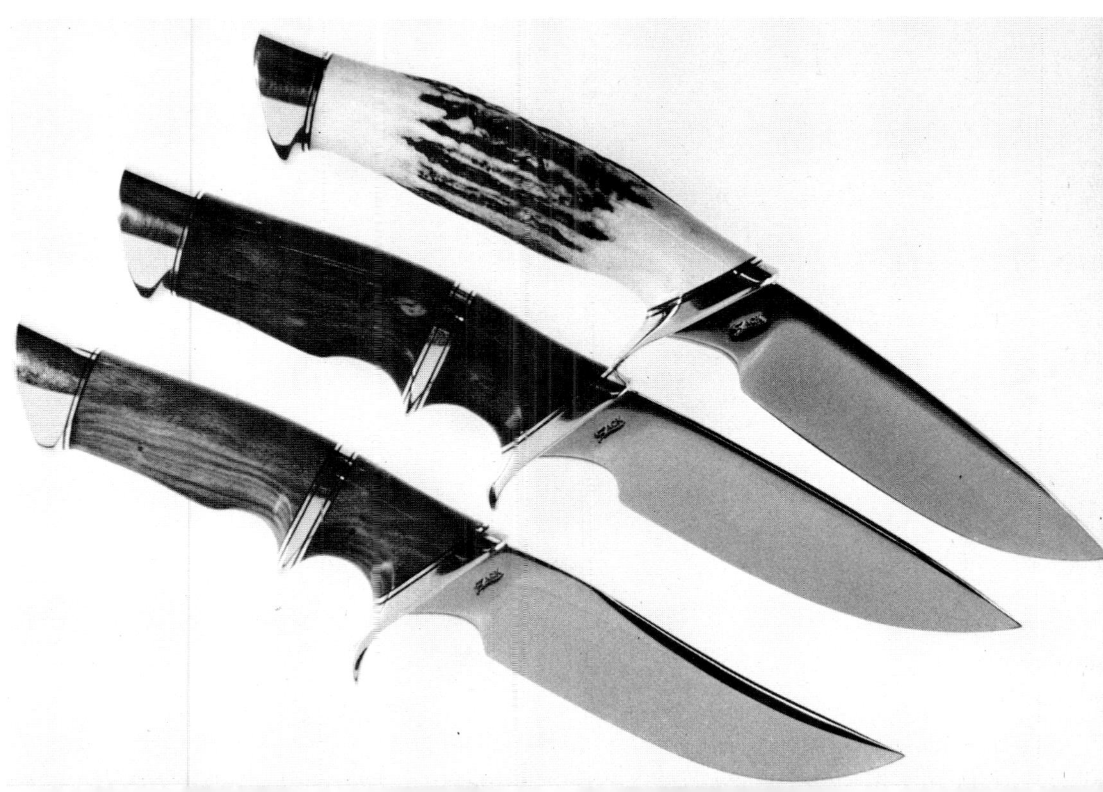

SHAW-LEIBOWITZ 76

Push daggers can be distinguished from each other by the shapes of their handles, but the principle remains the same for each — it is a weapon for concealment, and as such was favored by unscrupulous gamblers. The handle is gripped entirely by the hand, giving extreme force to the blow, which is often impossible to parry. The other three knives are variations on a theme, that of the shape of the handle. The decorative aspects, the acid etching, and the notches on the blade make these more collector's items than utilitarian objects. From top to bottom: a folding knife ('Single Bladed Folder'), a mixed model ('Kangaroo Folder'), and a skinning knife ('Rio Grande Skinner'). The handsome etching on the last one was executed, as usual, by a specialist. This knife is also striking for the complex structure of its handle. Nolen Knives, Mercedes/Corpus Christi, Texas

Right: These are fighting knives (or so they are supposed to be) rather than hunting knives. The blades are reminiscent of daggers or 'stilettos'. The one at the left is partially double-edged. Lew Booth, Boonton, New Jersey

All of these pieces commemorate the U.S. Bicentennial. The star attraction is the blade with the signatures of the signers of the Declaration of Independence. Lew Booth, Boonton, New Jersey

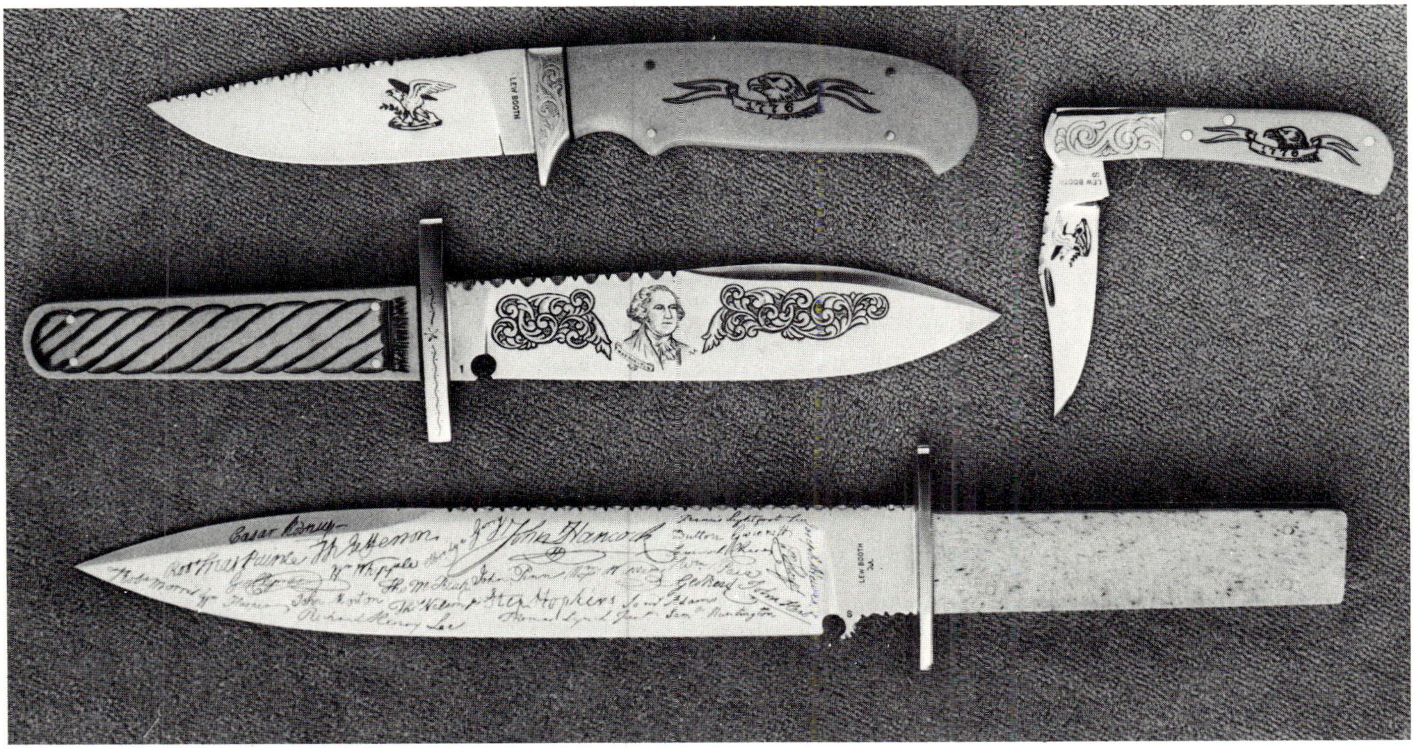

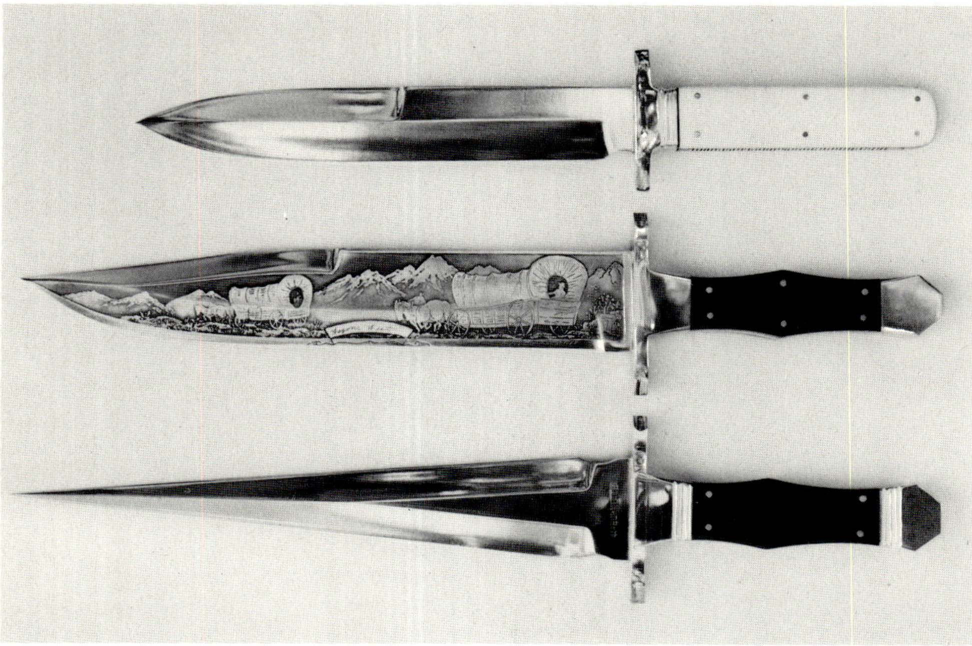

Left: The first of these three knives is named a 'Spearpoint Bowie': it is partly double-edged. The second is the 'Coffin Handled Bowie' and the last comes close to being an 'Arkansas Toothpick'. The smith and engraver (note the prairie schooner scene) are one and the same man. Harvey McBurnette, Piedmont, Alabama

Right: Although this dagger is honorific or a 'Presentation Dagger', nonetheless the blade is first class steel. It is the beauty of the handle, however, with semi-precious stones, nickel-silver ornamentation and silver wire inlay that really stands out. D'Alton Holder, Phoenix, Arizona

Below: This version of the 'Arkansas Toothpick' is the work of Zack, a professor of biology and mathematics and a knifemaker in his spare time, although he spends more time behind his grinding wheel than in a classroom. He is particularly interested in the side arms of the Renaissance and the methods by which they were made. Zack Knives, Pahokee, Florida

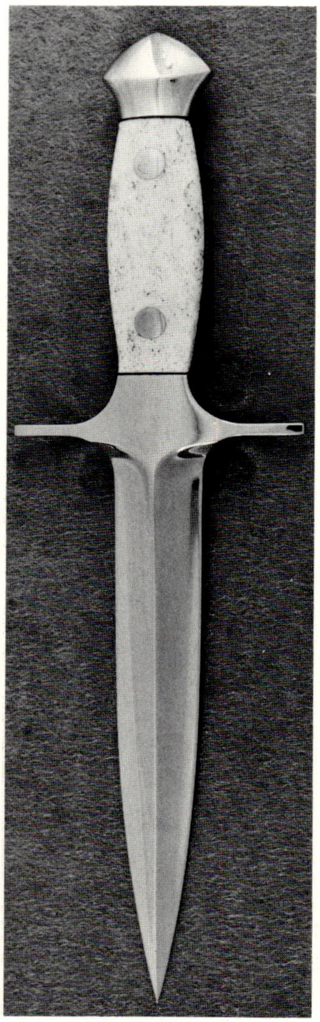

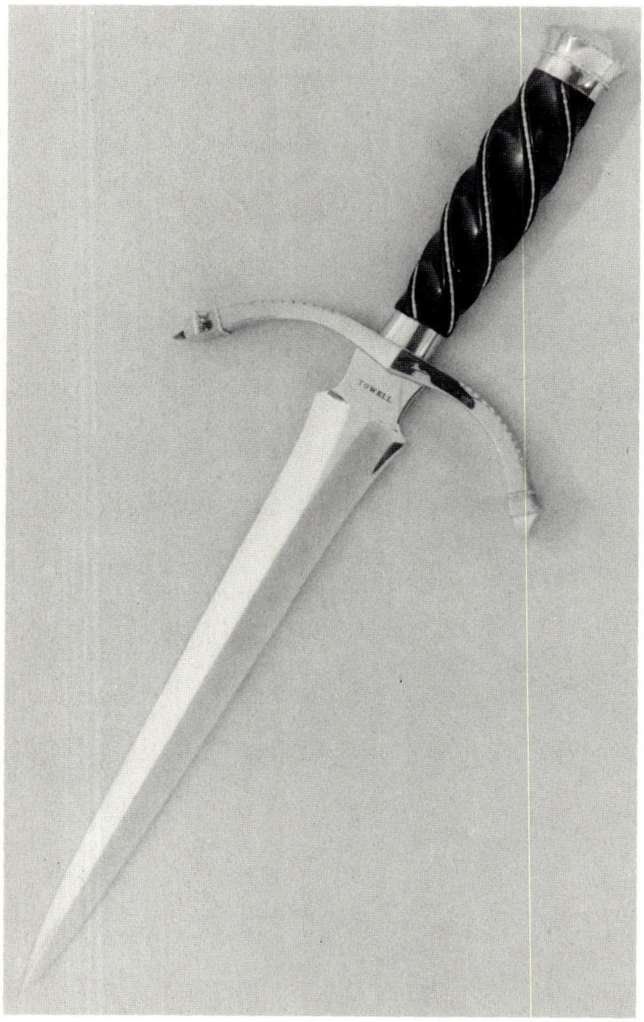

Far left: This 'stiletto' is 440-C steel. The handle is walrus ivory. Length of the handle is 3 inches; the weapon, 10 inches. Billy Race Imel, New Castle, Indiana

Center left: The 'Arkansas Toothpick' was a stiletto-dagger used by Confederate troops during the Civil War. The handle here is exotic wood inlaid with silver wire; the pommel is nickel-silver. Blade length: 9-1/4 inches. Dwight L. Towell, Midvale, Idaho

LEACH
KNIVES

The knife on the upper left is a hunting knife, on the right, a boot knife. The three large knives are, left to right: a 'Quannah Parker Bowie' (Quannah Parker was a famous Indian); the second is in effect a short sword; the third is a machete with a handle made of 'oosic' or walrus penis. The smaller knives at the bottom form a dagger-combat knife pair and are worn in the same walnut scabbard. The engravings are by Joe Rundell. Milo J. Leach, Swartz Creek, Michigan

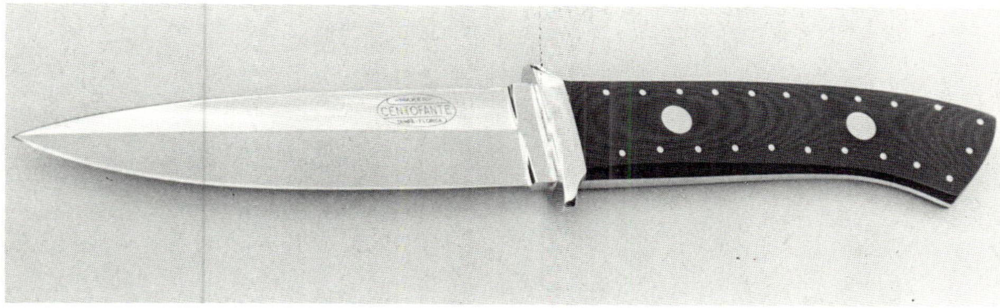

Three combat knives. The knife at the top right has a handle of Micarta and the rivets are nickel-silver. Length of the blade: 5 inches. Frank and Mark Centofante, Tampa, Florida — Below, the knife has a razor sharp edge and a walnut handle. (Blade 6 inches). D'Alton Holder, Phoenix, Arizona — Bottom right: knife with a brown Micarta handle. Note the similar guards. (Blade 7 inches). Dwight L. Towell, Midvale, Idaho

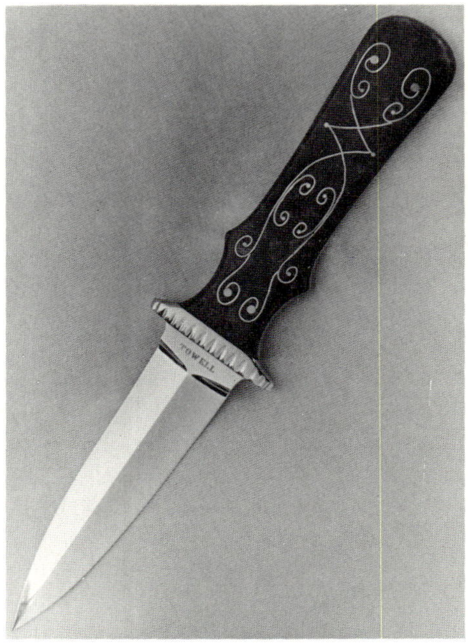

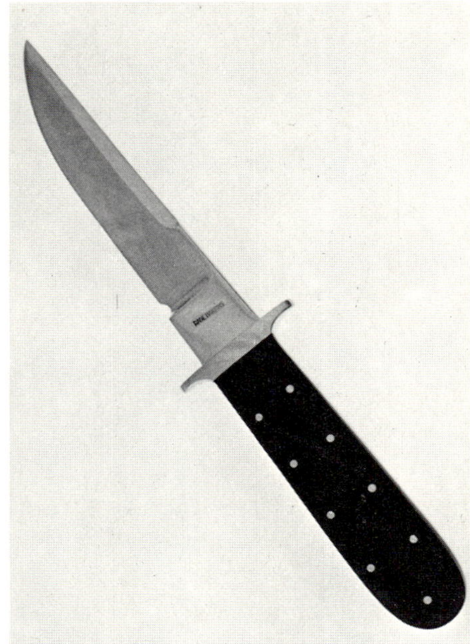

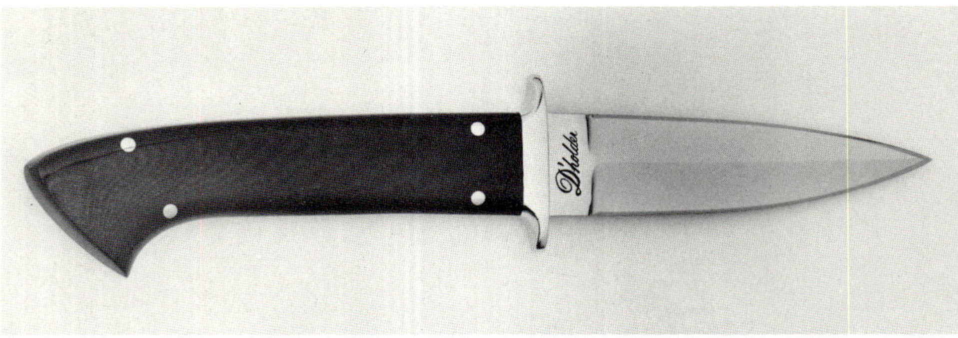

The knife which can fit into a boot has several advantages, the first of which is that it can be concealed. So there are boot knives, often double-edged (above, except right and below); 'boot dirks' (left) and boot daggers, sometimes with a single edge as the one above. But these terms are used loosely and can indicate one or another kind of knife. Blade lengths: 4; 4; 5; 4; 4 inches. Dwight L. Towell, Midvale, Idaho (the first two knives). Rade Hawkins, Red Oak. D'Alton Holder, Phoenix. Tommy Lee, Gaffney, South Carolina

These push daggers richly decorated with arabesques are of stainless steel. The rivetted handle grips are a continuation of the blade, and tang and handle echo the shape of the blade. The blade has the shape of a double triangle. Lew Booth, Boonton, Idaho

LEW BOOTH S

Right: Even the new synthetic materials can look very attractive. The handles of these two combat knives are Micarta, or plasticized cloth which can imitate various natural materials. In the first piece the basic material is a black linen and in the second a green canvas. The fittings are made of nickel-silver. Length of the blades: 4; 5 inches. Frank and Mark Centofante, Tampa, Florida

The blade of this 'Bowie' (9-1/2 inches) is of D-2 high chromium steel (RC 60 on the Rockwell scale). But this piece is, above all, a work of art. The handle is African hardwood and nickel-silver. The blade is chased with repoussé work. In all, the knife is made up of 53 parts! Harvey McBurnette, Piedmont, Alabama

All of these are throwing knives. Only the points are sharp; two knives do not have handles. The 'Artisten-Wurfmesser' is stainless steel; the others, in keeping with custom, are not. The one in the center is partly pre-oxidized; the others are varnished. Wingen, Solingen

CENTOFANTE

FROM THE KITCHEN TO THE TABLE

"It is precisely those days when I am alone at table that my dinner must be prepared with the utmost care: for then Lucullus dines with Lucullus."

Plutarch, *The Life of Lucullus*

To paraphrase Alexander Dumas, it can be said that without civilization, gastronomy — and with it the battery of kitchen implements and table utensils — would not exist. "The savage does not have to be induced to have an appetite", and he eats with his fingers.

The sacred character once attributed to food had its influence on *table knives*. The first *table knife* was usually also a knife for making sacrifices, before becoming the tool of a slave charged with the preparation of meats. In Greece and in Rome, where people ate reclining, the table service was limited to dishes for the meat and bowls for the drinks. Despite the prodigious eccentricities of Lucullus and the treatises of Athenaeus of Naucratis (*Deipnosophistes*) and Apicius (*De Re culinaria*), quantity came before quality. Even in the Middle Ages it was the practice to gorge on enormous quarters of meat and pies. For long, real dining rooms did not exist. Before the meal, servants brought in wooden panels and put them on trestles. For seats, benches (*banc* in French) were used, which explains the word banquet. Usually the lords and people of rank ate at a separate table. The setting — musicians, jugglers, incense — was more important than the food itself. Soon the napkin came into use — a great and necessary improvement since the fingers were used in eating. The meat, covered with sauce, was simply placed on a round of bread called a *tranchoir*, from which the dish known as a *trencher* was derived.

Trenching knives from that age, the 13th century, measure 11 to 12 inches long and usually about 2 inches wide. The knives were purchased by the steward and were kept under the supervision of the first carving squire, who carried them under his left arm, wrapped and wiped them before placing them on the table.

An anonymous French author wrote in the late Middle Ages that he "kissed the handles and set the two large knives before the place where the prince was to sit." He took care to "place the points away from the prince, covering them with the napkin, and to place the little knife between the two large ones." The kitchen instruments as well as those of the table were grouped under the general name *serving knives*; they included the double-edged *cutting knife* for the meats, *trenching-knife*, the *parepain* for cutting the *tranchoirs*, the *presenting knife* for serving the meat and the *chaplepain* for making bread crumbs for frying. The *convent table knives* carried notes of music, blessings and prayers on their blades. Although some monastic

This case for table knives is made of silver and is decorated with pseudo-Chinese reliefs which were once very much in vogue. Height of the case 6-3/4 inches. Nuremberg, beginning of the 18th century. City Museum, Sheffield

diets were indeed austere, it should be mentioned that the 'roots' and 'herbs' on their menus meant everything that grows in the soil and all green vegetables, often enriched with eggs, cheese and dairy products. Poultry could be eaten because, under the pretext that they were created the same day as the fishes, fowl were not always considered as meat. Dumay, the writer on gastronomy, has noted that it is not by coincidence that modern maps indicating well-known gourmet places bear an uncanny resemblance to earlier ones of convents and abbeys.

Keeping pace with the godly progress in the culinary arts came a refining of table manners. In the 15th century the luxurious habits of the elite contributed to the fortunes of the cutlers who were aided as well by the fashion that dictated the use of a different kind of knife for each occasion of feast. The handles were often decorated with inscriptions and inlays. Yet knives generally were rare, as indicated by Calviac, the chronicler writing in 1560: "The Italians boast that each has his own knife. But the Germans find that this is a singular attribute...: they become annoyed if one is used in front of them. On the other hand, a whole table of Frenchmen may be obliged to share two or three knives."

Although the spoon was a contemporary of the knife, the fork, born in the 11th century, did not come into its own until the reign of Henri IV — along with the plate, the napkin and the chair. Cardinal Pierre Damien considered scandalous the way in which Princess Marie, wife of Jean Orseolo, used to eat: "Advancing her disregard of customs as far as not using her fingers to eat with, she used for this purpose golden forks which were presented to her by slaves." The napkin and long handled fork were intended to spare the large ruffs at the wrists of the costumes worn by men.

In the 17th century, use of the knife became firmly established at table. A wooden handle replaced that of ivory used during the Renaissance. Two knives of this epoch which belonged to Gaspard de Coligny, Duke of Châtillon, have been preserved. Another, with two blades (of which one is silver), has a handle chased with carved silver in the form of flowers and birds: it bears the inscription in old French *Mon cœur volle d'amour* (My heart soars with love).

The first question to ask before judging the quality of a kitchen knife is whether it is the right knife for the job. There are many kinds of knives, differing in shape of the blade, length and width, the size of the point, whether or not it has teeth, the kind of handle, etc. From left to right are: a kitchen knife, a peeler, a tomato-peeler, a steak knife, a cheese and

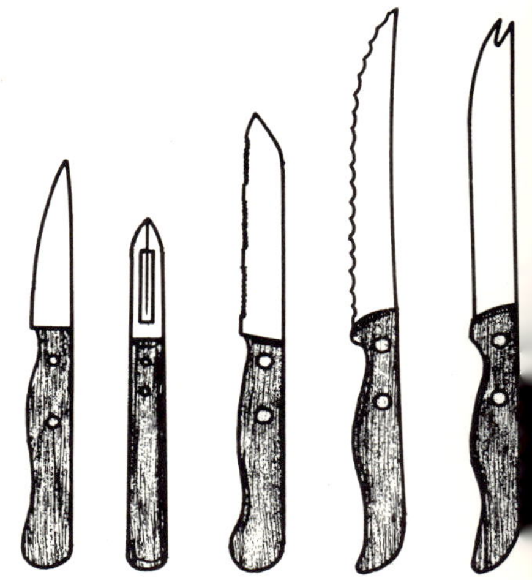

The art of gastronomy

The reader will have noticed that up until now we have not spoken of *kitchen knives*. However, from the 18th century on, the age of quality replaced that of quantity. The Regent of France, later Louis XV, the favorites of the court and then the grand ladies, all began to amuse themselves with cooking. It was the epoch when 'mahonnaise', (now mayonnaise), and bechamel sauce were invented and the battery of kitchen utensils grew. J.-J. Perret, the 18th century cutler, distinguishes among the following knives: 'Firstly, the *classic kitchen knife* with a strong edge for slicing meat and cutting small bones (poultry wings and legs, etc). The *tranchelard*, which must have a fine edge and a thin spine, used for cutting bacon and for preparing lardons or larding strips; it can be bent for better cutting the strips by holding the handle in one hand and the point in the other. The *belt-knife* is a semi-tranchelard; the cook uses it for preparing meat. The *peeling-knife* exists in different sizes and is worn at the belt... as are most of the others.

Even among knives of the same type, there can be some differences of appearance. Here for example, is a range of chef's knives of different shapes and sizes. The length of the blades varies from 3 to 10 inches; they are made of stainless steel. The handles are made of pressed wood especially designed for dish washers. Zwilling, Solingen

tomato knife, a lemon knife, a grapefruit knife, a decorating knife, a scaler, a rabbit bleeder, an oyster knife with hand guard, a roast knife, a carving and boning knife, a bread knife, a bread and paté knife, a Swedish bread knife, a leg-of-lamb knife, a chef's knife, and a ham knife. From the catalogue of the Etablissements Thérias, Thiers, France

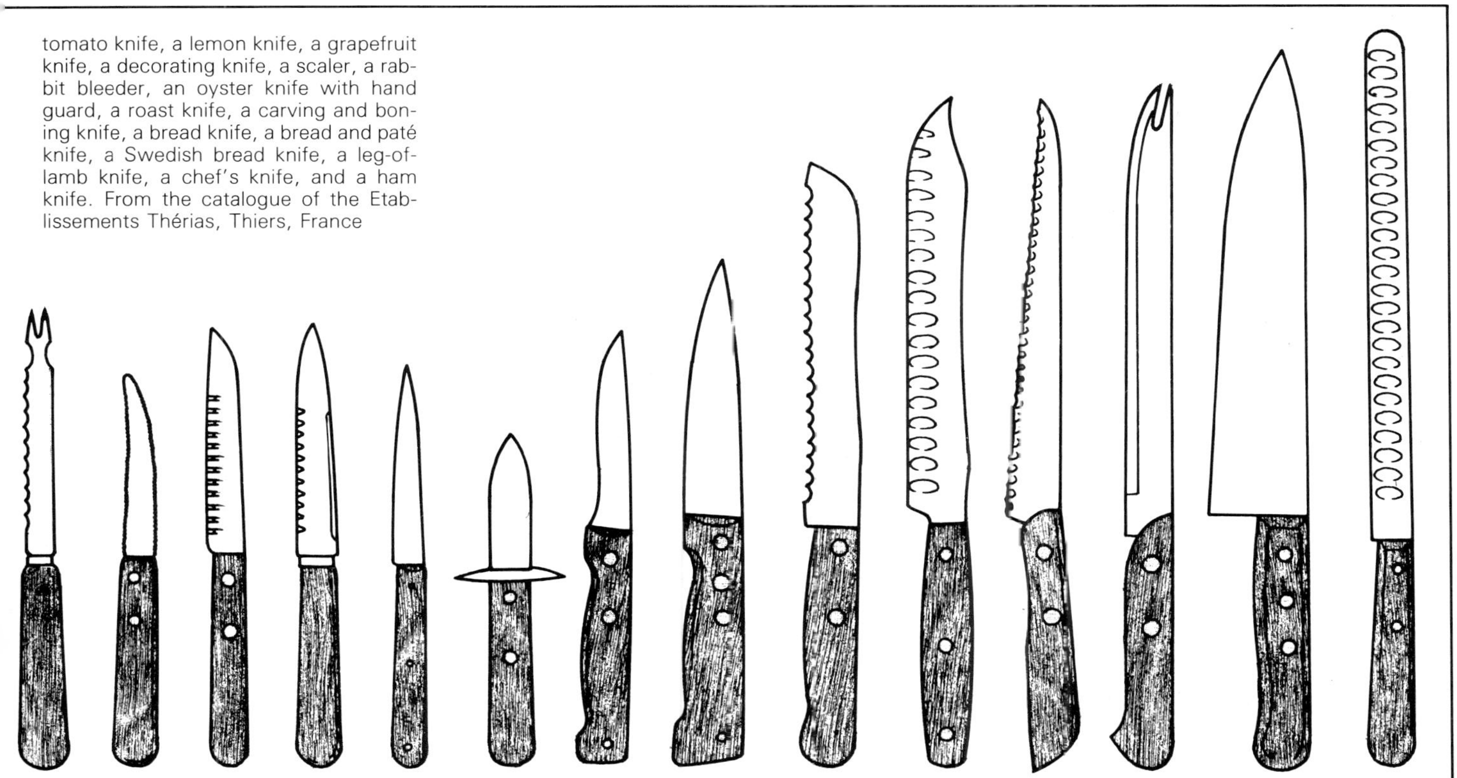

Left to right: Two vegetable knives, followed by a decorating knife, a tomato knife, a meat fork, a general purpose knife, a carving knife, a bread knife, a meat knife, and a frozen food knife. Length of the blades 2-3/4; 3-1/4; 2-1/4; 4; 4-3/4; 6-1/4; 8; 8; 7-1/4; 8 inches.

The notches on these knives are purely decorative. These kitchen knives are similar to those previously described, with the slight difference in that the cutting edge is very bevelled. Length of the blades 2-3/8; 3; 3-1/8; 5; 6-1/8; 6-3/4; 8; 9-1/8 inches, Wingen, Solingen

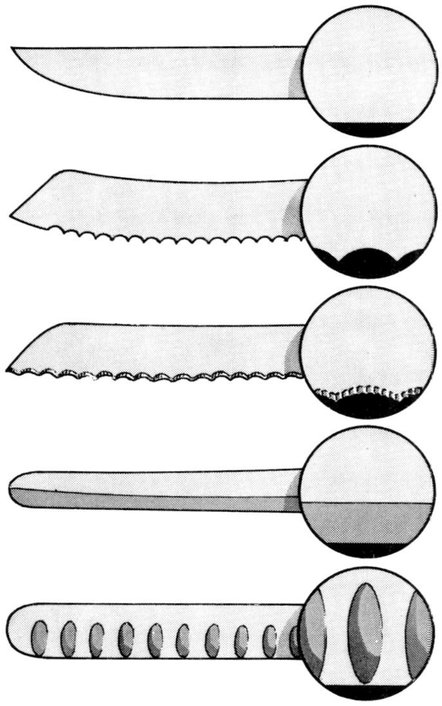

Pantry and kitchen knives, like all knives, can be distinguished according to the kind of edge they have. Here are five typical cutting edges. The first is the simplest and it can be found on professional kitchen knives as well as peelers, carving knives or on chef's knives. The teeth of saw blade knives as, for example, bread knives, have hollowed out scallops to which can be added fine vertical grooves. This is often the case for blades for frozen food knives. According to the type of edge, most blades must be more or less bevelled but it can happen that the bevel takes up half of the width of the blade. Among kitchen knives this is often the case with ham knives. The scalloped blade can also be found on ham and galantine knives and some bread knives. The edge is more or less pronounced according to the type of knife. Based on the catalogue of the Etablissements Thérias, Thiers

The master chef has a wooden scabbard which is made for holding from two to five knives."

It has often been said that the style of table services has followed that of furniture, yet in the time of Louis XV *table knives* were very simple in shape, in contrast with the rococo which reigned in the decorative arts. *Table knives* were embellished with taste and the noblest of materials were used in their making. The handles were made of ebony inlaid with ivory, silver, or mother-of-pearl. Some were also made of tortoiseshell or mother-of-pearl with line, groove or bow-shaped inlays of gold or silver. The bolster rings and their escutcheon-plates were made of the same precious materials. Again, in the 18th century knives were sometimes made with interchangeable blades, one of steel and another of gold or silver. In 1764 the cutler Gavet was authorized to use a drop-press to stamp out silver handles — thus lowering the price. Knives with porcelain handles were also extravagantly prized, and brought fabulous prices. The reign of Louis XVI saw the fashion of 'drapery' handles, which resembled flounced cloth.

The Directory and Empire periods saw flat knives with escutcheon plates fixed to the bolsters upon which were stamped coats-of-arms, and these came to be called 'French knives'. Under the Restoration, stamped fittings replaced the bolster rings and line-decorated escutcheon plates. Even knives with silver handles took on this form. Sometimes the handles of the latter were shortened so that an additional material could be used in the embellishment of the handle. Under the reign of Charles X the escutcheon-plates were no longer used; an engraved palmette constituted the end of the handle. In 1827, the *fulcrum blade* was invented; its base protrudes and keeps the blade from touching and soiling the napkin when placed on the table. This invention is considered to have marked the beginning of the modern *table knife*. From that time on the design was generally more sober, albeit with interludes of decorative exuberance. The blade itself, pointed and outward swelling in shape during the 18th century, became parallel with a rounded end. In the present day it is difficult to determine such a general trend, all the more so since the materials used influence the shape of the knife. Scandinavian design, one of the first to really affirm itself after the *art déco* style, has now been displaced by even simpler and above all, more functional, forms. Finally, with electroplating and mass-produced cutlery, silverplated and stainless steel knives have come on the market in large numbers. Fashion hesitates between copies of classical styles and the creations of modern designers.

Knives have become more numerous and diversified: *vegetable knives*, *peeling knives*, *oyster knives*, *ham knives*, *carving knives*, *bread*, *butter*, *cheese*, *fruit*, *cake* and *ice-cream knives*. The variety of forms and functions is unending.

How to choose a knife

A good knife is above all one that is made and adapted for the use to which it is put: a *carving knife* should not be used for slicing bread, nor a *fruit knife* for meat.
The quality of a knife depends above all on the choice of materials and the way it is made. In traditional craft methods the blade would be forged into

the desired shape starting from a blank or bar by heating and hammering the red-hot steel on an anvil; more often, however, the blade is ground out of the blank. This latter is called 'stock removal'. What distinguishes craft methods or 'benchmade' knives from factory production is the absence of the use of moulds and the simple assembly of pre-fabricated parts; but even on the production line, hand labor still plays an important rôle and this system should not be confused with typical mass production. Assembly can be carried out on an assembly line basis. A high-quality piece of metal, or blank which for an ordinary knife should be hard, flexible, and resistant to chemical agents, is cut roughly to the desired shape or shaped with a power-hammer. Tempering and the 'drawing of the temper' or 'return' are the next steps, which impart certain qualities to the steel. Grinding involves several successive operations, beginning with the use of rough abrasives to reduce the blade to the thickness and the edge desired. Repeated sharpenings are followed by polishings using finer and finer abrasives. By this time the handle has been ground and polished. Many manufacturers put on an edge by grinding, and the final sharpening is done by hand. Some manufacturers polish the edges, making them more tapered and usually less durable.

Factory-made knives are those upon which the least number of manual operations have been performed and, consequently, they cost the least. But many manufacturers take short-cuts when it comes to the quality of the materials, the steel, rivets and so forth. For example, some steels contain nothing more than iron and carbon. As a result they are not very sharp and are brittle. A sharp edge may be put on such blades without difficulty, but they will not keep it; they are subject to rust, and bend easily. Another handicap of mass-produced knives is the lack of quality control in their manufacture. It is much easier to spot a defect and reject an unsuitable knife at an ear-

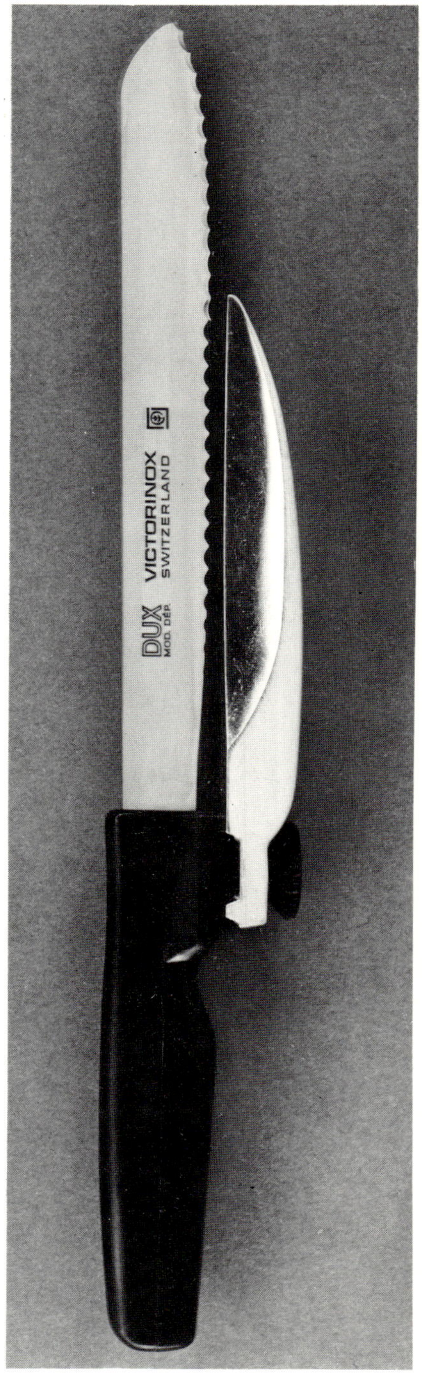

This 'Dux' knife by Victorinox has a ruler that guides the knife to cut uniform slices of bread. Swiss made. Coutellerie du Mont-Blanc Collection, Geneva

Frozen foods are deadly to the edge of an ordinary knife. This led to the creation of a special kind of knife with a steel that is hard without being brittle, and with fine, sharp teeth. In the knife pictured, the handle is of plastic. Zwilling, Solingen

There are several different-looking cheese knives. Some end with a small hood to spear pieces of cheese in order to serve them. The model above may seem a bit specialized but is in fact well adapted to cutting a large round cheese. The one on the right is called an all-purpose knife, and can be used for dough, butter and all semi-soft substances as well as fresh cheese. Actually, it is more of a spatula than a knife in the strict sense. Zwilling, Solingen

ly stage of the manufacturing process, when at least some of the work is carried out by hand. In addition, mass-produced knives can easily be distinguished from other kinds by the often depressing uniformity and banality of the models.

The strongest handle is one that is ground and machined out of the same piece of steel as the blade. The two can never come apart and the handle will be exempt from cracking and deformation. But the heat-conducting properties of metal make it impractical for use with hot food. On the other hand, it has a definite advantage in that it can be washed in a dishwashing machine. For those who like wood handles, hardwood is preferable because it is less likely to burn, split, or become deformed, while fir or pine is more sensitive to heat and absorbs water. The advantage of wood is that it is not heat-conducting and is less slippery, even when wet or greasy. However, it is not resistant to shocks and fares less well in a dishwasher. Moreover, wood provides a hospitable surface for the growth of bacteria. It is recommended that the blade and handle be washed and dried separately. And finally, if knives with wooden handles are to be properly cared for, they should not be allowed to soak in water.

Handles made out of synthetic materials have become much more popular in recent years. The plastics are not perfect — some of them are not even very strong — but, on the other hand, they are moderate heat conductors. Vinyl is a good example of a quality plastic: it is particularly useful for knives intended for fish and shellfish since it is impervious to salt water. For *kitchen knives* plastic is less than ideal because it burns or melts too easily. As it is slippery, a ridged handle is preferable to any other type. Plastified wood is a good synthetic which has the appearance of wood and the advantage of plastic; in addition, its less porous surface resists bacteria growth.

In brief, when buying a knife, the decision should be made between an attractive knife which is perhaps not well tempered, or something less dazzling but more durable. If a heavy duty knife is needed, the choice becomes more crucial: between a bevelled edge blade which, without a doubt, will become dull, but which can be easily resharpened, or a super-keen hollow ground blade which will also inevitably lose its edge, but will lose it for ever. In selecting *kitchen knives*, a positive point to look for is the use of rivets for holding the blade and handle together.

A word about the care of knives: all knives become dull. Sharpening the blade with a grindstone should be done at least once a year, but it is best left to a specialist. But to whet a knife edge it is enough to pass each side of the entire blade three or four times over a butcher's steel, and there are other methods as well. It goes without saying that a steel should never be used to sharpen saw-edged or scalloped knives.

The knife on the left is an all-purpose kitchen knife but it would be a mistake to use it to cut bread because the crust would ruin the edge. For this, the knife on the right should be used.

The art of carving

An art — for it is an art — which brings together the culinary world and that of cutlery, is the art of carving meat. In a restaurant the carving of a dish is a responsibility left to the head-waiter; it requires good taste, since the food must be presented in a manner which is both aesthetic and appetizing, and it requires culinary knowledge, because the carving should bring out the dish's particular qualities. The operation must be carried out quickly so the

More and more home cooks are returning to sturdy old-fashioned looking knives. From this point of view nothing is better than wood for the handle. A quality blade is the result of a dozen operations starting from a thick bar of steel which has been straightened, polished, ground, and sharpened evenly on both sides. On the right, on the wall is a meat fork, a 'butcher's steel', two kitchen knives with slightly bevelled edges, a boning knife, a carving knife, a slicing knife; then from top to bottom, two chef's knives, a ham knife, a bread knife, a cheese knife, a serving knife with rounded tip, an oyster knife with hand guard, a peeler and an oyster knife without guard. Fonderies de Cousances, Ancerville, France

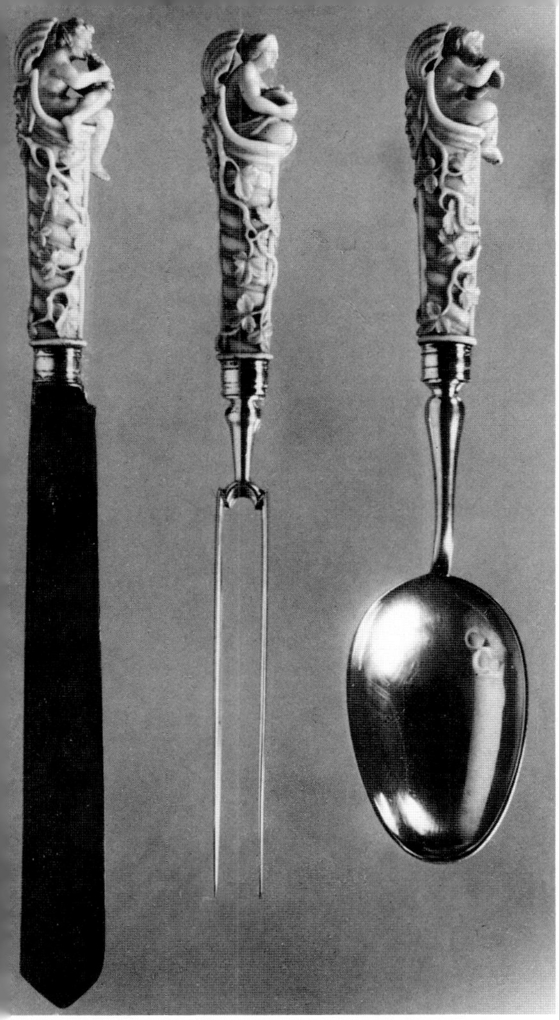
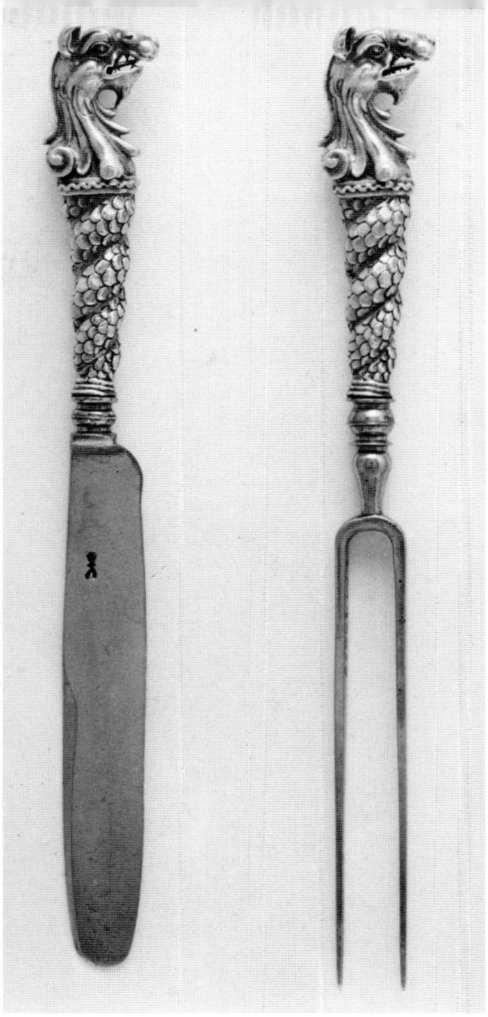

The first picture shows a German table service of the 18th century decorated with Bacchic scenes. City Museum, Sheffield — In the middle are a knife and fork with carved handles that terminate in a stylized head of a lion. Perhaps by Hans Konrad Oeri, Zurich, around 1670. Schweizerisches Nationalmuseum, Zurich — The S-shaped piece on the right is doubtless German and can be dated from the same period. Museum für Kunst und Gewerbe, Hamburg

Left: These two series of knives are German and date from the 17th century. Each handle (four to five inches high) is a representation of a Biblical figure: above, the Blessed Virgin and five Apostles. The series below includes the fall of man (third from left) but may contain non-religious scenes. The figures are in ivory. City Museum, Sheffield

food does not get cold. Each dish calls for a different cut: roast beef should be cut into thin slices with just a little fat, while tongue or galantine should be cut in rather thick, transversal slices. Also, one should know that roast beef fillet must be turned over on the plate in order to be cut into slanting slices, and that ham is supposed to be cut with a long, flexible and well-sharpened knife.

Books specialized in the subject give a multitude of details concerning the carving of various dishes, how the knife must be held, at what angle, and so forth. There are nine principal types of kitchen cutting operations, besides peeling. Drawing is the preparation of poultry and wildfowl for cooking by removing the skin or feathers and the giblets. Cleaning is the removal of scales, gut and gills from fish. For these operations a *chef's knife* is used — or, for fish, a *fillet knife*. Boning is one of the most difficult operations — in addition to the knives named above, it usually involves a special knife, according to what type of animal is being boned.

For cutting raw meat either a *slicing knife* or *chef's knife* can be used, or perhaps both in succession. The pieces have to be cut in clean strokes. In the case of beefsteaks or scallops the direction of the cut is very important. Preparing cold cuts, especially in the case of poultry and lamb, is a specialty in itself, and the list of other kinds of cutting — dicing, trimming, chopping, as in the case of vegetables and herbs — can continue up to decorative trimming, particularly in the case of fruits and certain vegetables such as artichoke hearts and tomatoes.

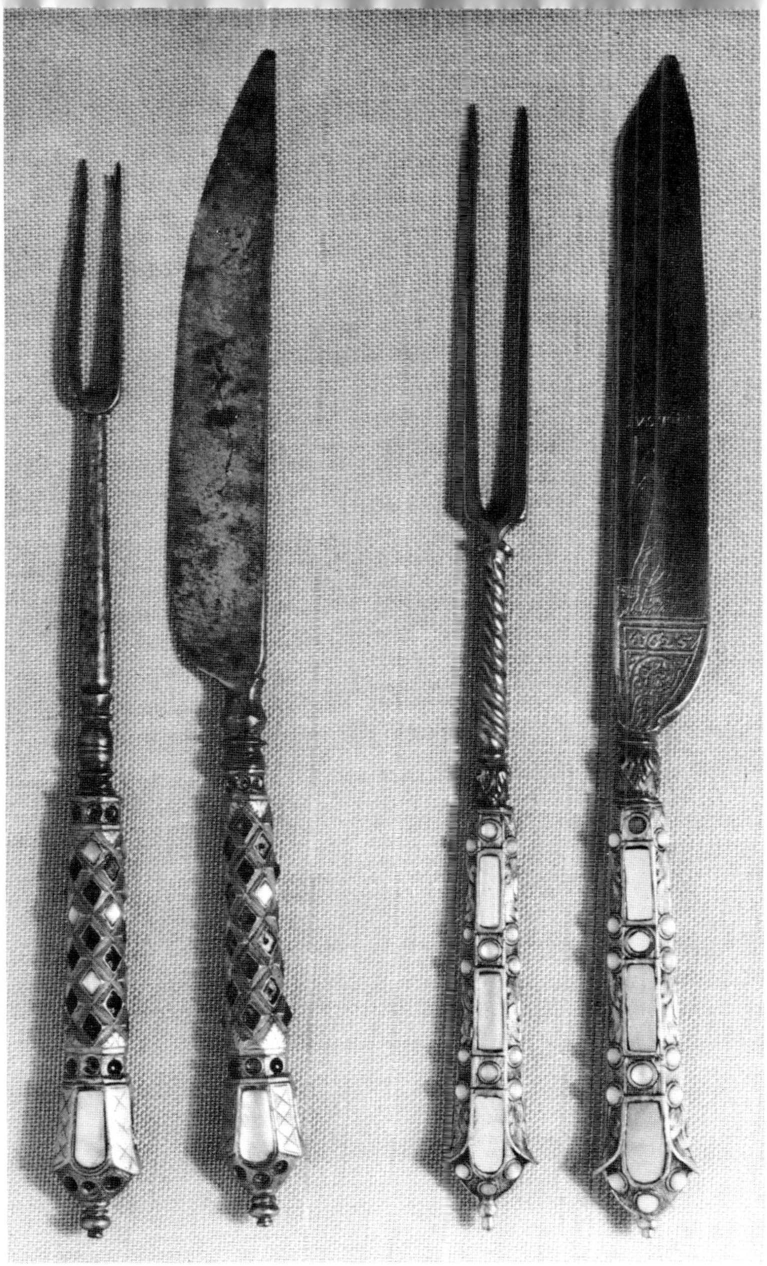

Left: Germany, and the city of Solingen in particular, have been famous since at least the 14th century for their knives and side arms. The Solingen Museum is well-known for its blades, whatever their use or origin. These admirable table services date from the 17th to the 19th centuries, some having saxony porcelain handles, others handles of silver, wood, semi-precious stones and even crystal. Deutsches Klingenmuseum, Solingen

These two pairs are French and date from the 17th century; the second carries the date '1625'. The handles are gilded metal inlaid with pearls, and, on the left, semi-precious stones Length of the knives: 5-1/8 inches. City Museum, Sheffield

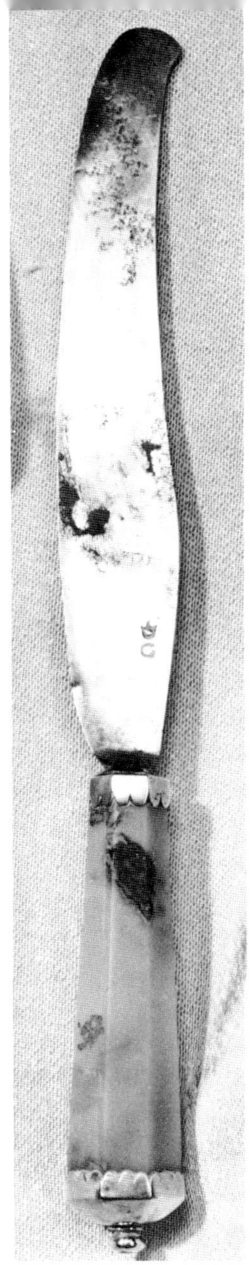

This knife, with a chased silver handle and steel blade, bears the cipher of Henry III. It thus belongs to the end of the 16th century and is of French manufacture. Musée Bouilhet-Christofle, Saint-Denis

This handsome knife has an agate handle and fittings of silver. France, about 1650. Musée Bouilhet-Christofle, Saint-Denis

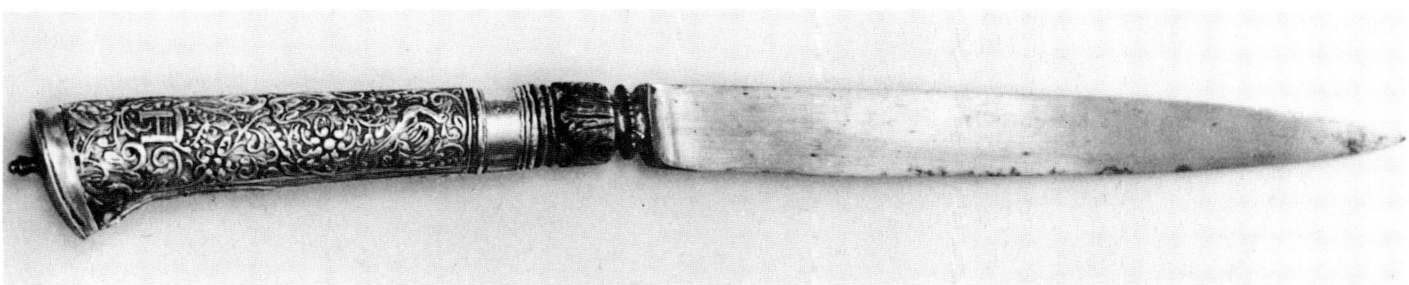

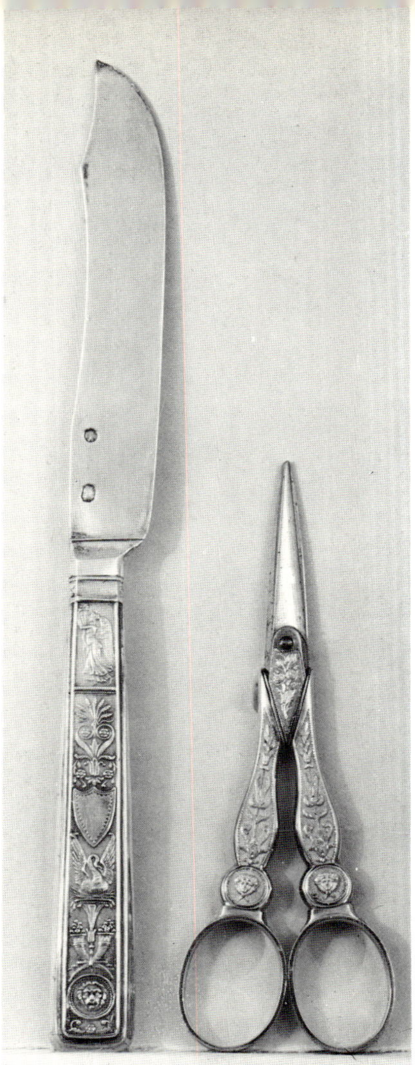

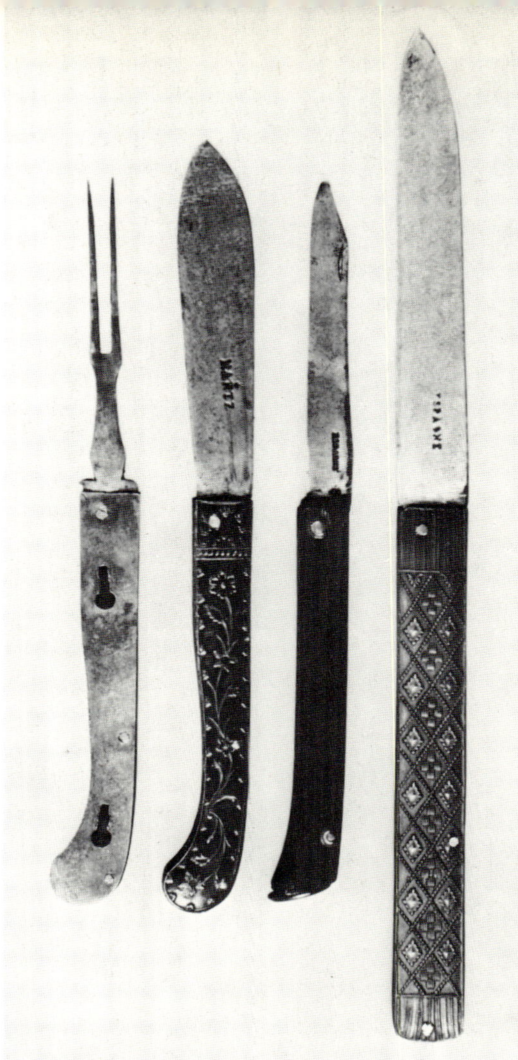

Of these two sets the second appears to be older than the first, and yet the opposite is true. The table service dates from the 18th century. The knife and scissors on the left, in gilded silver, are German and probably date from the end of the 17th century. Schloss Nymphenburg, Munich — The pieces of the service on the right fold up, and the fork, though in a decrepit state, reveals how the handle is assembled. These four pieces probably come from Sheffield. City Museum, Sheffield

These English table services are works of art. The first spoon at left, in gilded silver, has an agate handle. Daniel Treadwell, London, around 1675. The following knife was part of a set; it is a wedding knife with an amber and ivory handle. John Jenkes, London, 1633. The second pair was originally made up of knives but one of them was modified into a fork about 1660. The handle is agate. John or Jonas Melser, London. The handles of the third pair are ivory painted green and red inlaid with silver wire. Swain, around 1690. The fourth pair with painted 'pistol' handles, dates from the end of the 18th century. Finally, the handles of the last pair are ivory and tortoise shell; London, about 1710. Note the rounded and oversized ends of the last three blades, a common feature of 17th and 18th century knives. City Museum, Sheffield

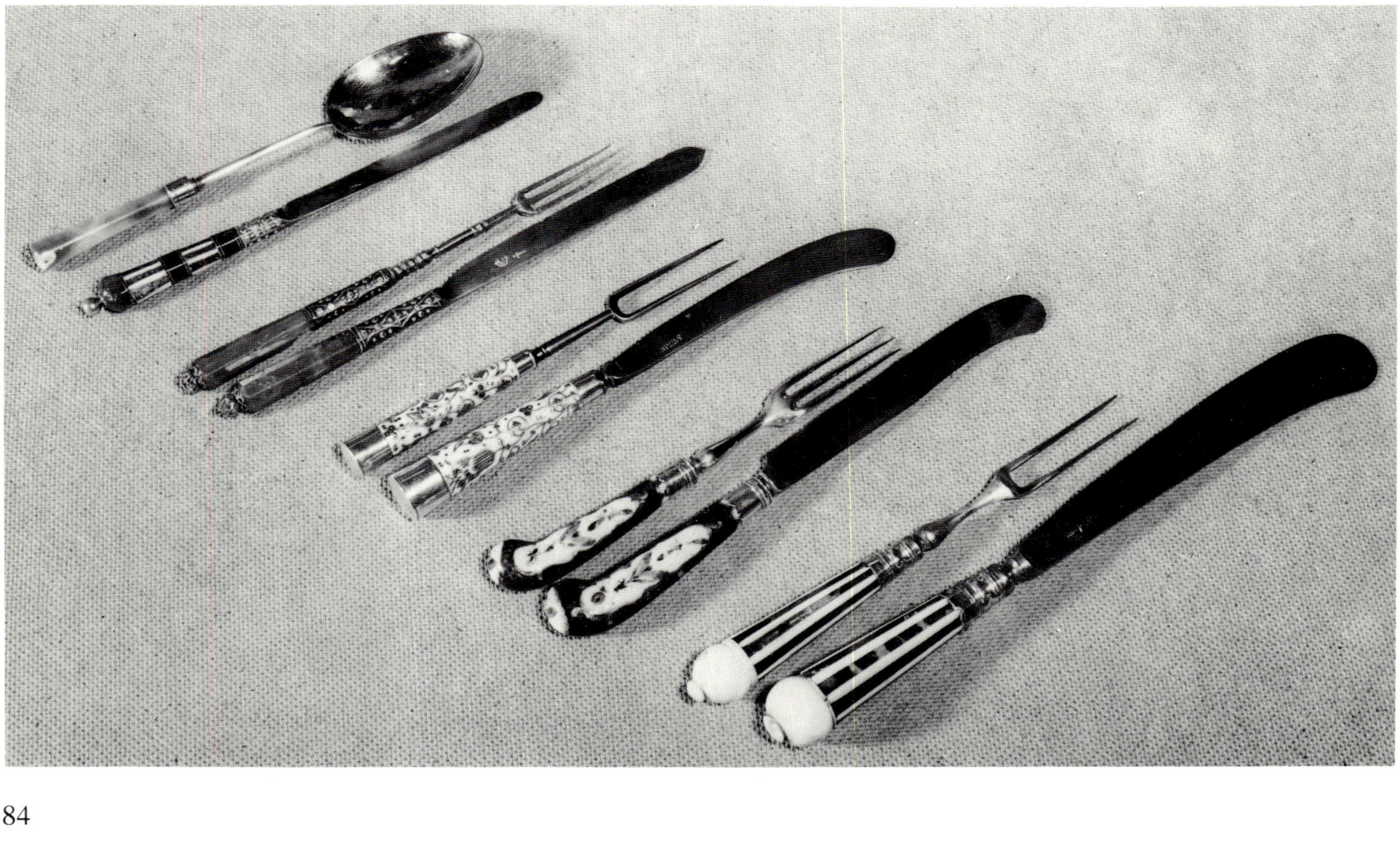

All of these knives have silver handles. The one at the left is from Augsburg (beginning of the 18th century); those following are French (end of the 17th century; about 1720; beginning of the 18th century). The third has double scallop shells at the butt and the bolster, as well as a fulcrum hilt. Musée Bouilhet-Christofle, Saint-Denis

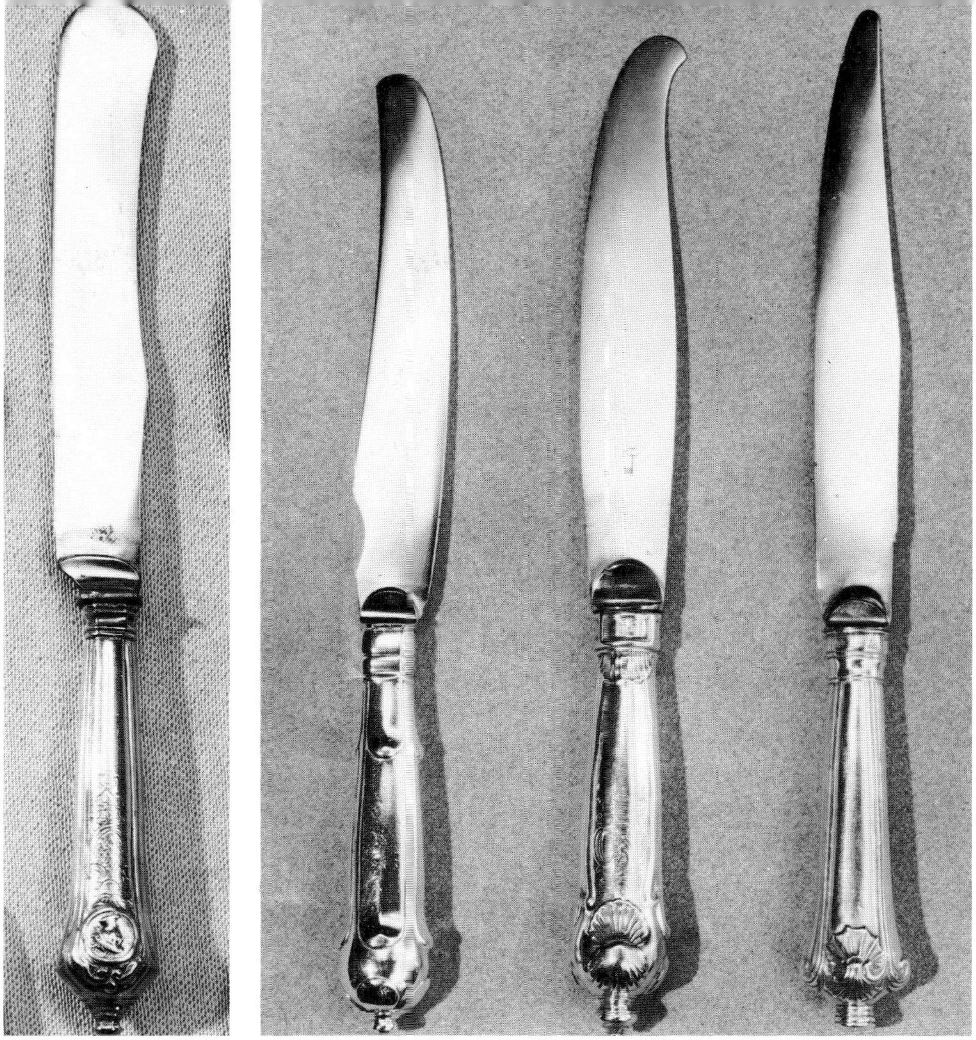

The use of contrasting materials can produce marvelous effects, as shown by this ebony handle inlaid with silver. This knife is of Strasburg origin, 1714. Musée Bouilhet-Christofle, Saint-Denis

The first two of these three pairs date from the beginning of the 18th century and the last pair from the middle of the 18th. All of the handles of these German-made table sets are ivory. The lengths of the pieces are from around seven to nine inches. City Museum, Sheffield

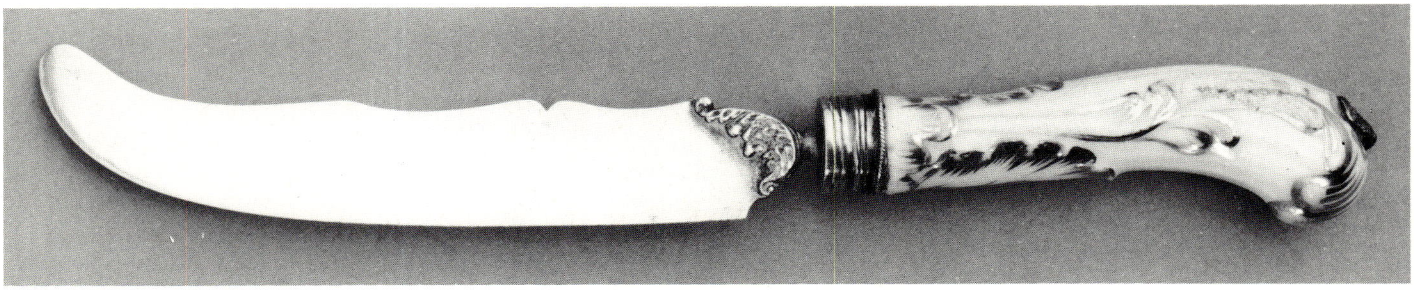

This German knife dates from the 18th century. The bolster and trimmings are silver. The blade and the handle are saxony porcelain with superbly twisted shapes. Musée Bouilhet-Christofle, Saint-Denis

Right: Increasing travel, combined with the lack of table services, led to the growing use of travelling sets. This one bears the arms of Pope Pius VI and dates from the 18th century. The spoons and forks are silver, as are the handles of the knives, which continue along the steel blades. Musée des arts décoratifs, Paris

Like the knife above, this table service is of German origin and dates from the second half of the 18th century. It comes from Nuremberg. Spoon and fork are silver, as is also the 'pistol' handle of the knife. Residenz-Museum, Munich

Note the striking contrast between these two knives, the one widening towards the base, the other flaring out towards the end, both of which are rounded curiously. The first three pieces date from about 1770; the last two are also 18th century. The handles on the left are pewter inlaid with green diamonds. Sheffield. The ones on the right are millefiori. England, 18th century. Length of the knives: 11 inches. City Museum, Sheffield

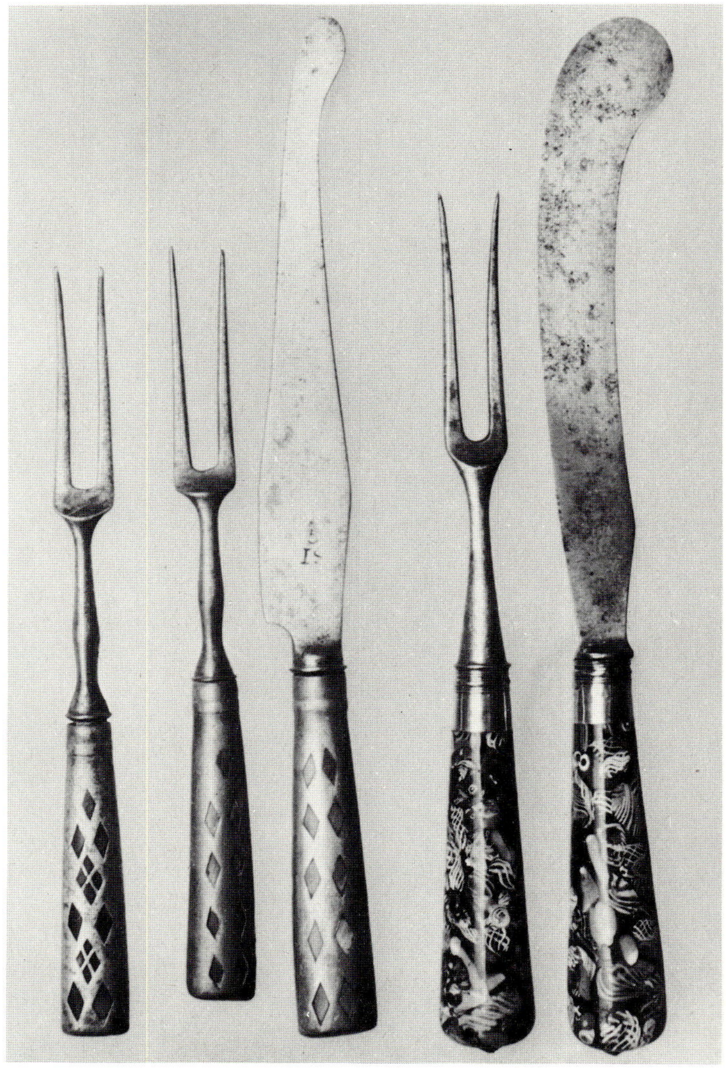

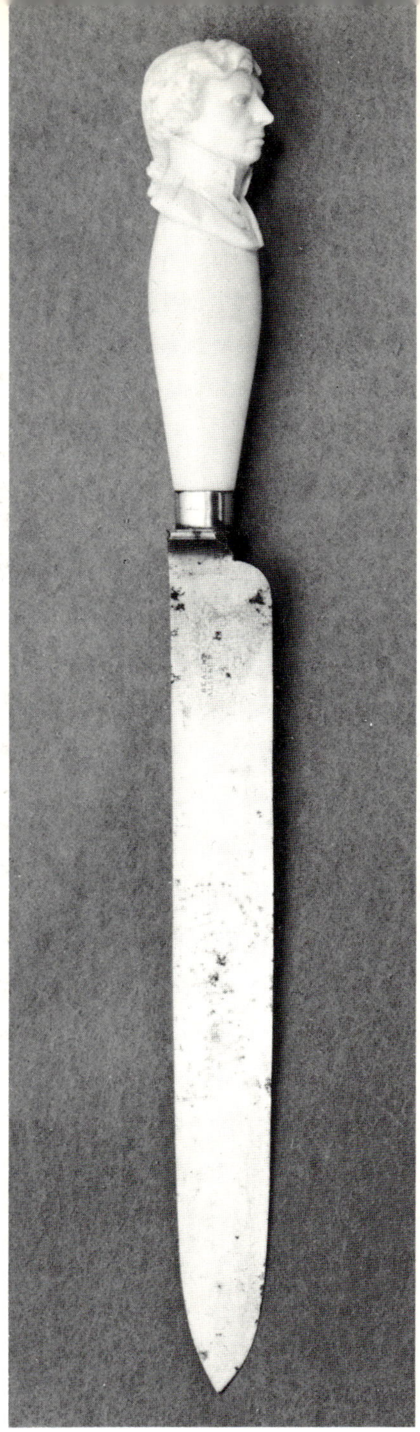

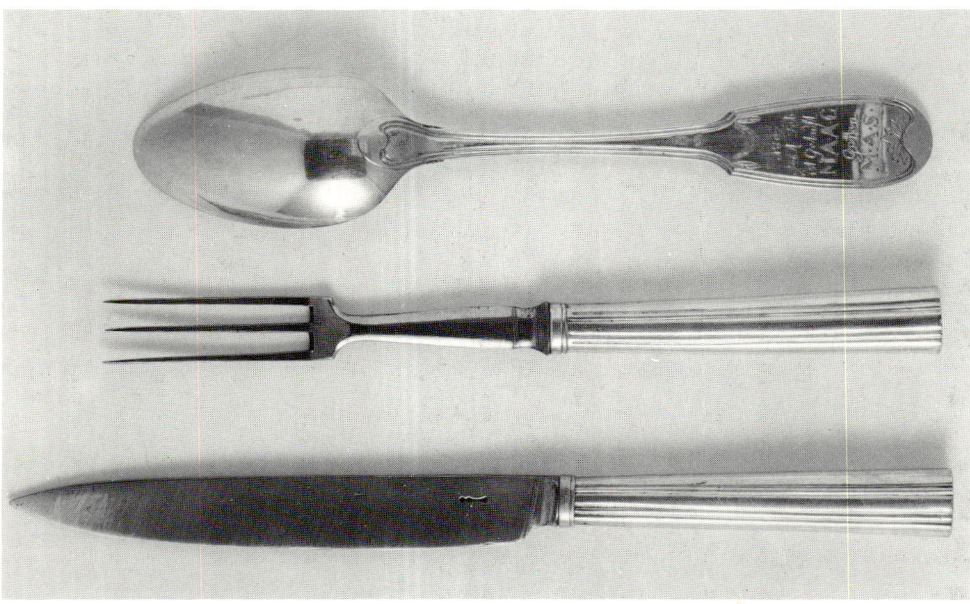

Above, left: The outside of this case is leather chased with gold, the interior chamois leather. The blades are removable and can be placed separately into the case. The handles are lathe-turned ivory. Height of the case: 5-5/8 inches; diameter: 3-3/16 inches. Austrian, probably 1779. City Museum, Sheffield

Above: This carving knife is part of a set. The ivory handle is surmounted by a carved head of Lord Nelson. Sheffield, 1806. Length: 18 inches. City Museum, Sheffield

The handles of this communion service are fluted silver. The knife has a sharpened point. Jakob Müller, 1807. Schweizerisches Nationalmuseum, Zurich

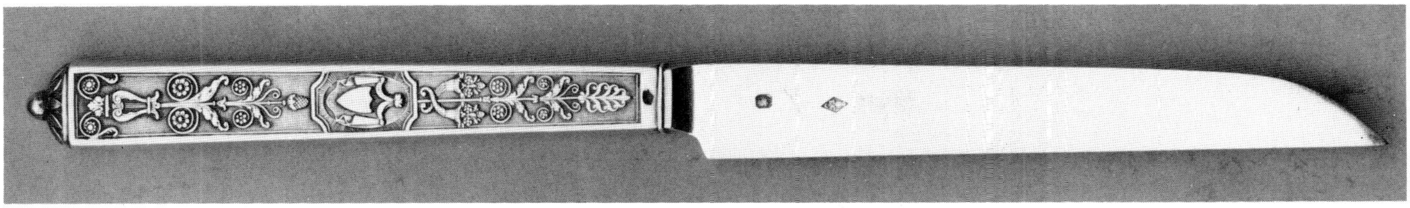

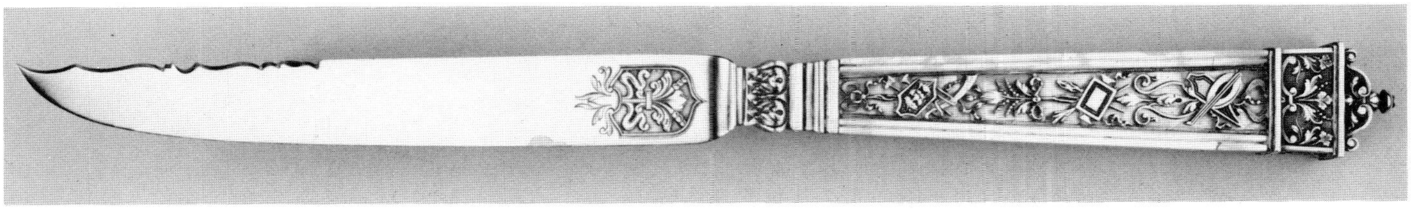

These three French knives (above and bottom left) are First Empire, Restoration, and Second Empire. The first is silver-gilt. The one below, created by Christofle in 1855 for Napoleon III, is silver-plate. The handle of the second is inlaid ivory. Musée Bouilhet-Christofle, Saint-Denis

Table service made out of silver-gilt with a case. All of the pieces can be folded. Northern Germany, 1820. Museum für Kunst und Gewerbe, Hamburg

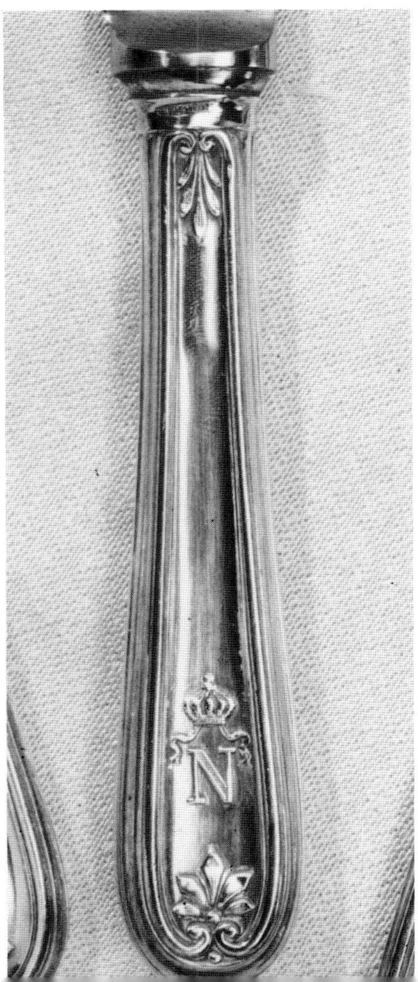

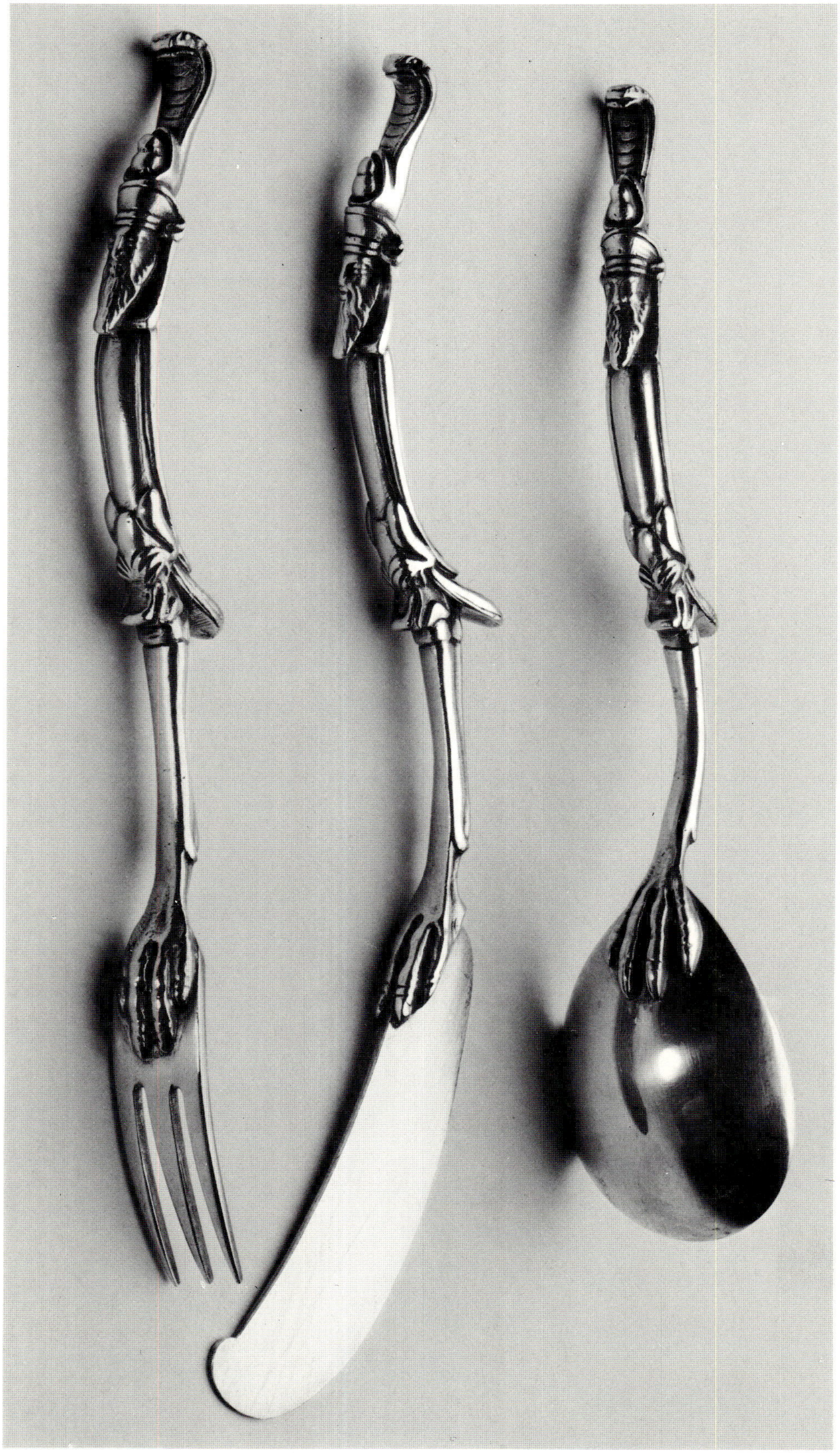

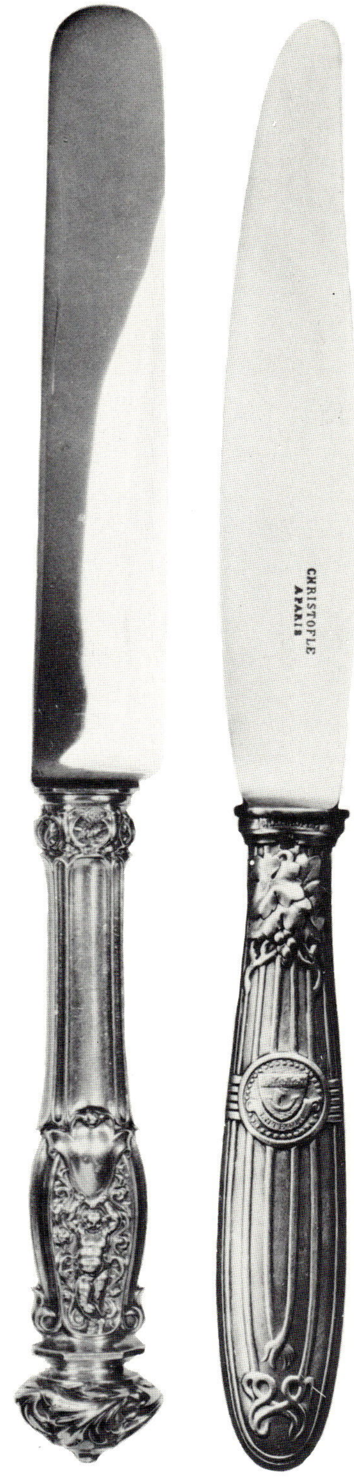

Above, left: This knife by Froment Meurice, in silver (1840) was designed by Dupouchel and executed by Klagmann. Musée des arts décoratifs, Paris — Above, right: The handle of this knife by Christofle, his 'Lutetia' pattern, is silver-plated. End of the 19th century. Musée Bouilhet-Christofle, Saint-Denis

Left: The exuberance of Art Nouveau is here inspired by the Gallo-Roman treasure of Hildesheim. Musée Bouilhet-Christofle, Saint-Denis

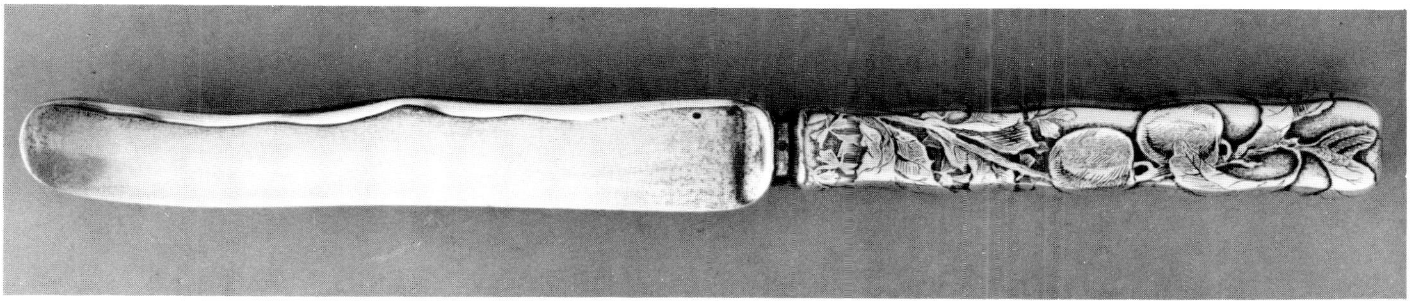

This knife (1900) is by Tiffany. This famous American jeweler was the first in the west to use mocoumé. Musée Bouilhet-Christofle, Saint-Denis

Right: This dessert spoon and fish service was created by Henry van de Velde in 1902. It was manufactured from 1903 on by Theodor Müller in Weimar. Museum für Kunst und Gewerbe, Hamburg

Below left: This silverplated model is called the 'Atlantic' and was created by Gio Ponti for Christofle. Musée Bouilhet-Christofle, Saint-Denis

Below center: This model, called the 'Atlas', has a silver handle and was designed for Christofle by Lanel especially for the ocean liner 'Normandie'. Cardheilac Collection. Musée Bouilhet-Christofle, Saint-Denis

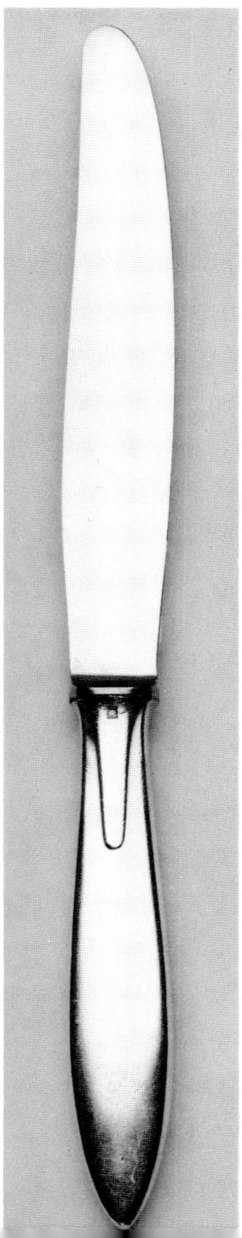

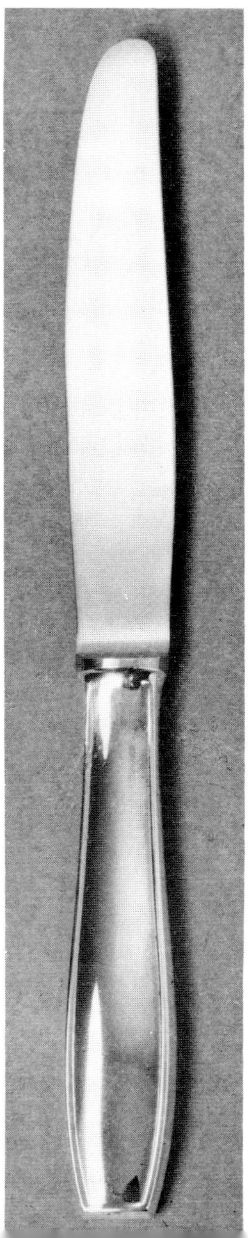

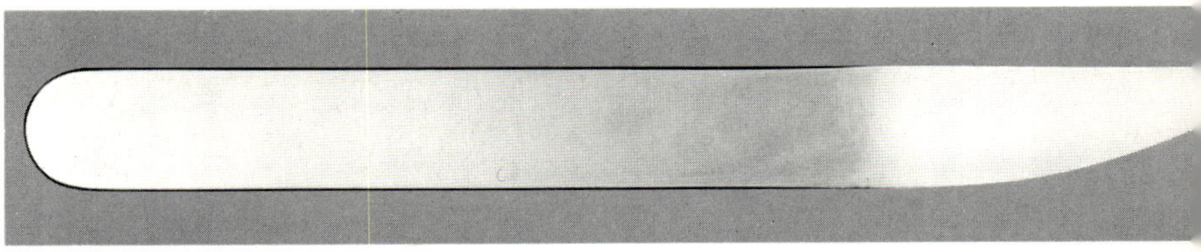

Above: This prototype 'Christofle', of an extreme simplicity, is the work of Henri Bouilhet. It was never manufactured. It is silver-plated. Musée Bouilhet-Christofle, Saint-Denis

Left: This set was created by Fjerdingstad for Christofle in 1925. The handle of the knife is ivory. The twisted handles of the fork and spoon go well with the simple knife handle. Musée Bouilhet-Christofle, Saint-Denis

Right: This 'Homage to Giacometti' was made by Gio Ponti exclusively for Christofle. The tormented stringlike aspect of the Swiss sculptor's work is clearly recognizable. Of course it is questionable how well these pieces would go on a dining table... but that is another story. Musée Bouilhet-Christofle, Saint-Denis

'Chinese Ivory Cutlery'. Despite its name, it is not made of ivory, although it resembles it. The blade and the rivet are of stainless steel, and the handle of synthetic resin. Note the symmetry between the blade and the handle. David Mellor, Sheffield

This 'Java ivory cutlery' set is a good example of certain very sober styles current today. The blade is stainless steel, the handle synthetic resin. David Mellor, Sheffield

1
2
3
4
5
6
7
8
9
10
11
12
13
14
15
16
17
18
The Charmer
Koken

FOLDING KNIVES

"With his knife he cut the hard bread... and with his knife he scraped rotten fruit... and with his knife he cut walking sticks... it's with his knife that he carried out all of the labors of life".

Anatole France, *The Wicker Dummy*

This illustration shows a selection of folding knives of the 18th and 19th centuries. 1 and 2: The knife in the shape of an alligator and the other in the shape of a lion, are both from Sheffield and date from the middle of the 19th century. Length closed: 2-1/2 inches. 3: The grips or covers are white wedgewood imitation of jasper, ornamented with green and mauve motifs. Staffordshire, end of the 18th century; 3-1/8 inches. 4 and 5: These two knives of imitation agate are of the same period and origin as the preceding one. Length: 3-1/4 and 3-1/2 inches. 6: English again, this piece has covers of real agate, 18th century; 3-1/2 inches. 7: The covers, decorated with a relief, are a conglomerate of horn. Sheffield, beginning of the 19th century; 3-1/8 inches. 8: In this example, tortoise shell has been inlaid with mother-of-pearl. Sheffield, 19th century; 2-5/8 inches. 9: These staghunters are inlaid in a conglomerate of buffalo horn. The knife was intended for export to Mexico. Sheffield, 19th century; 4-1/2 inches. 10: The handle of this model, which is in the shape of a boot, is made of tortoise shell. Sheffield, 19th century; 2-3/4 inches. 11 and 12: A knife and fork pair. The grips are enameled. France (?), 18th century; 3-1/4 inches. 13 and 14: Two German razors. The grips are aluminum decorated with Art Nouveau scenes. Late 19th century; 5-1/4 inches. 15: This knife of German origin is made of stamped aluminum. The design shows a factory. Late 19th century; 3-3/8 inches. 16 and 17: The grips or covers of these two knives are engraved mother-of-pearl. Early 19th century; 3 inches. 18: Brass, decorated with silver filigree. Sheffield, early 19th century; 3 inches. City Museum, Sheffield

A few years ago, before *pocketknives* were relegated to a category similar to costume jewelry, knick-knacks, or tourist souvenirs, it would have been difficult to imagine a man, be he a beggar or a businessman, a ne'er-do-well or the local parson, leaving his house without his *pocketknife*. Many people, at one time or another, have felt embarrassed because they had forgotten their knives... or because they had none. Today, however, the *pocketknife* seems to be making a comeback.

The advantage of the *folding knife* is that it can be carried with less of the danger inherent in an edged instrument. That is doubtless why *folding knives* have existed since ancient times and why, in the Middle Ages when so many people had to carry their own knives, the use of the *folding knife* spread widely.

The *folding knife* (which is not always a *pocketknife* — some are quite large) springs from the simple idea of the handle also serving as the sheath for the blade. This metal case is composed of liners parallel to each other and held together by rivets; the spring which makes up one side also holds the blade open or closed. The blade is usually single-edged, can be pointed or rounded, has a hole at the smaller end and can pivot about a 180° axis.

The most widespread *folding knives* of the Renaissance were apparently *oyster knives*. Folding *pastry knives* have also been found. The 18th century produced handsome models, some with springs and some with pegs, that is, without springs. Among the latter, one of the most famous on the continent, perhaps because it was the cheapest, was called the *jambette*. Also known as a *eustache*, after its inventor Eustache du Bois, a cutler of St. Etienne, it is still made today. The blade is fixed on a single peg so that, when closed, the point disappears into the handle in the front. At the back it is retained by a small protrusion forged at the heel of the blade. The *capuchin knife* had two pegs; the blade pivoted around one, while the other held it once the knife was open.

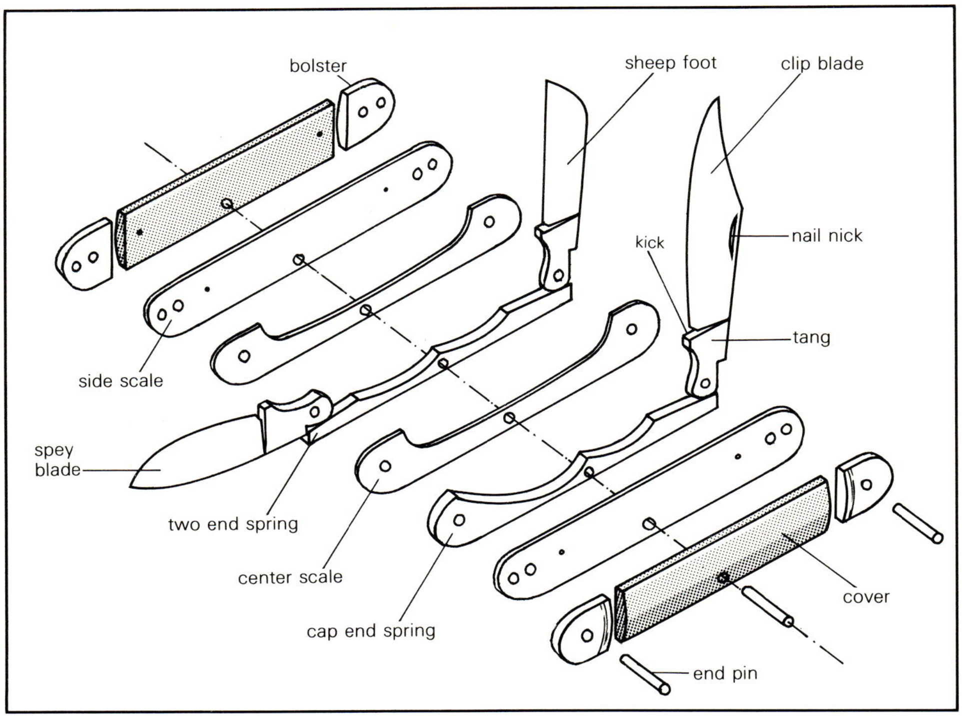

This diagram shows the various parts of a typical folding knife. For a diagram of a knife with a safety catch refer to page 107.

Spring folding knives were more numerous. One of them was the *crooked knife*, which had a spring and three pegs; its name came from the form of the handle. Soldered to its scales, the *military knife* had a trimming on top and an escutcheon plate below, both of silver. In the *double-bladed knife,* each end of the handle presented a blade. There are many other varieties.

The *springless knife* or *Berge knife*, named after the cutler who created it, had two blades placed on the same tang. They were independent of each other and could open and close either together or separately — turning like a drawing compass. The knife without pins was also sometimes called an *eagle's head*. It was mounted on studs hidden by the covers; from the outside no pins could be seen.

One kind of knife which was very successful during the 18th century was the *secret knife*. As its name indicates, a special procedure known only to the manufacturer and the owner could open it. In practice, the 'secret' was often very simple. In the *groove knife* for example, the hole in the tang of the blade was oblong; the end of the handle had a groove. When the knife was closed a spring moved the blade into this groove. To open the knife it was necessary to slide the blade on the tang pin. The *button knife*, of which there were several kinds, was based on the following principle: the tang had a little stud which would come to rest on a spring; to close it the spring had to be pulled back. In the knife with a sliding rosette, one of the decorative rosettes, which looked like all the others, moved a spring with a hooked end which fitted into a notch in the tang, opening the knife.

Among unusual, not to say spectacular knives, were the folding *bayonet-knives* made for hunters, which, as the name implies, could be used at a rifle end.

The principle of this French 'secret knife' is reminiscent of a safe: it opens when both hands point to six o'clock. The handle is horn and brass. Early 18th century. Total length: 11 inches. City Museum, Sheffield

Multi-blade and other knives

Folding knives have the advantage of being able to carry several different blades on the same handle. Knives with two blades did not become common until the 18th century. Then, from two the number of blades increased rapidly to three and six. There was no reason to stop. In the last century some knives had fifty, sixty-five, even a hundred blades. These knives were often heavy, clumsy, and more remarkable as curios than as practical tools. Besides, they were expensive.

In the meantime various accessories were added, starting with the corkscrew. For the record, here is a list of the blades and accessories of a knife shown at the Paris International Exhibition of 1878:

blades in steel and silver
scalpel
veterinary lancet
three kinds of penknives
scraper
paper-cutter
flower-cutter
string-cutter
cigar-cutter
ivy-cutter
budding-knife
pruning-knife
scissors
saw
ruler
two corn-cutters
various toilet instruments:
  nail cleaner, tooth-pick, ear cleaner,
  tweezers for dental tools,
  tweezers for splinters
string threader
ribbon threader
pipe cleaner
various tools:
  spatula, stamp,
  two gouges, gimlet,
  chisel, file, sculptor's file,
  screw-driver, punch
buttonhook
corkscrew
needle
grafting-knife

Among the folding knives of the 19th century was the *couteau à la d'Estaing*, which appeared at the time of Louis XVI and was named after an admiral of that name. The following description comes from Camille Pagé, the

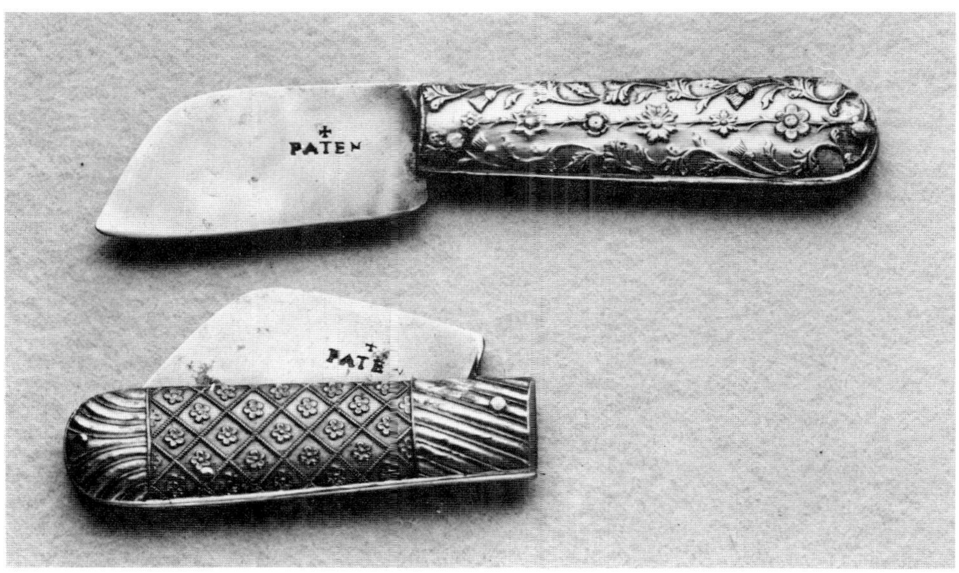

These two spring knives are the work of George Paten, cutler of Sheffield. The handle is embossed brass. The blade folds only partly into the handle. Later part of the 18th century. Length closed: 3 inches. City Museum, Sheffield

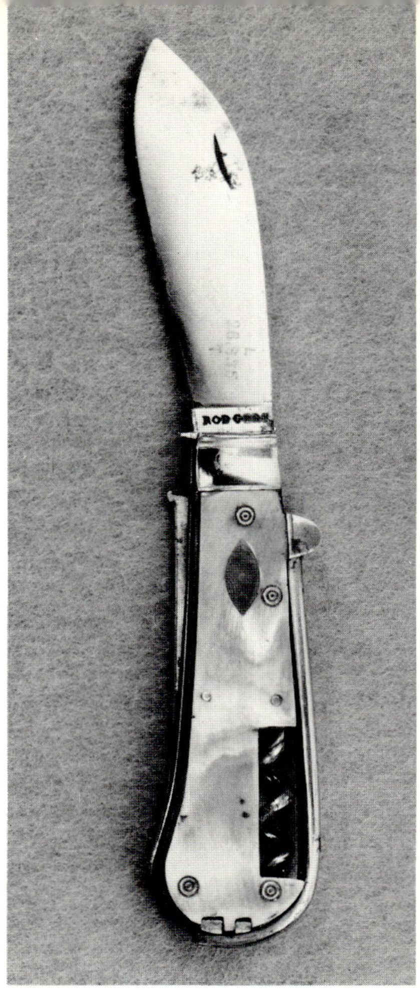

Above: This folding knife with a spring contains various accessories, including a corkscrew. The covers are horn. Joseph Rodgers & Sons, Sheffield. Late 18th century. 3-1/2 inches. City Museum, Sheffield

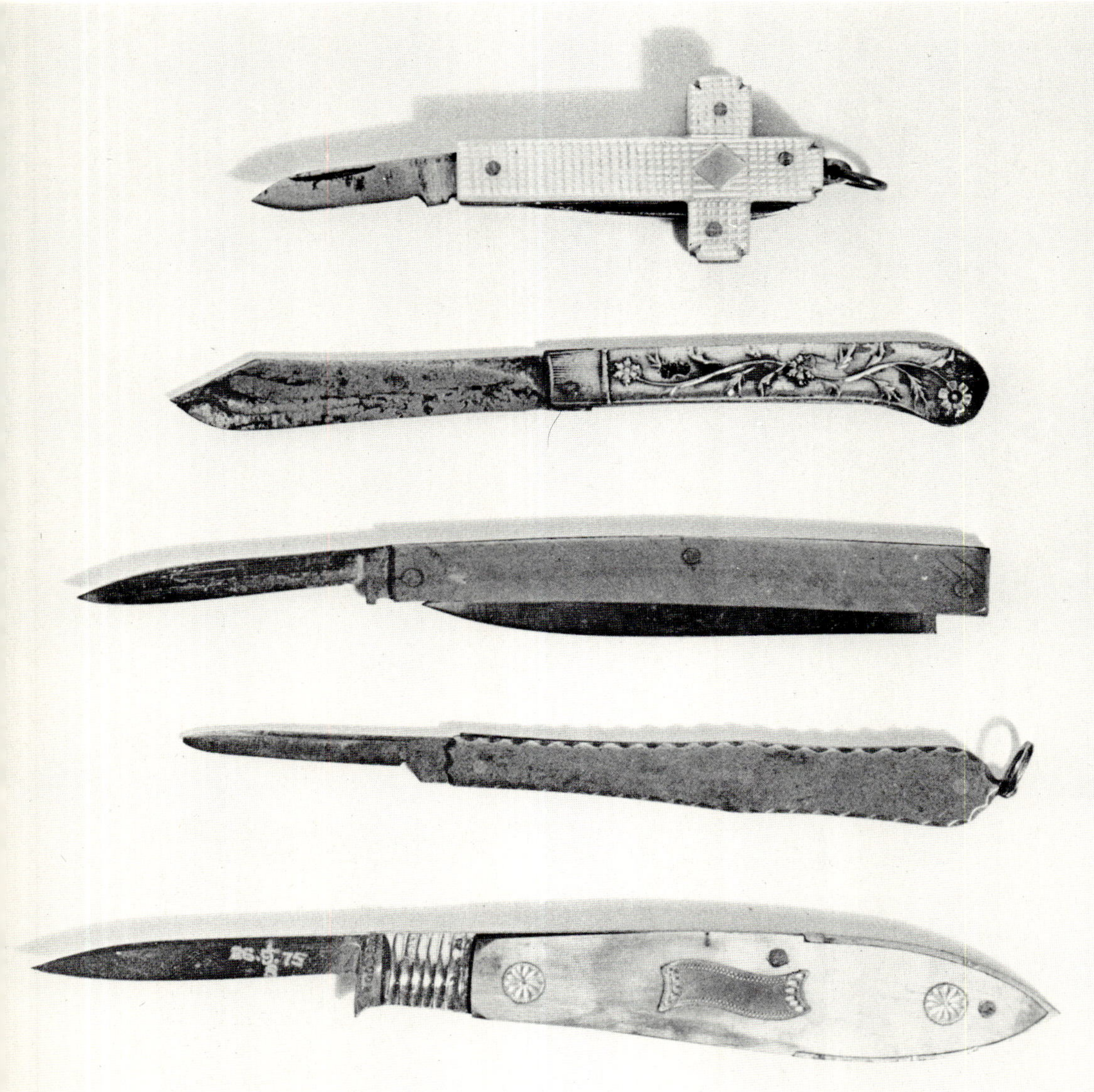

Left: Folding knives took on odd shapes, such as the knife pictured here at the top, with two blades, in the form of a cross. The grips or covers are engraved mother-of-pearl. Probably Sheffield, 19th century. Length closed: 1-5/8 inches. The covers of the knife below it are stamped brass. Late 18th century. 2-3/4 inches. The following model, with two blades, is luxurious with covers of green tinted ivory. Joseph Rodgers & Sons, Sheffield. Late 18th century. 2-3/4 inches. The fourth piece has steel covers. England, end of the 18th or beginning of the 19th century. 2-1/2 inches. Finally, the covers of the last one are horn inlaid with silver. Joseph Rodgers & Sons, Sheffield. End of the 18th century or beginning of the 19th. 3-1/4 inches. All of these are penknives. City Museum, Sheffield

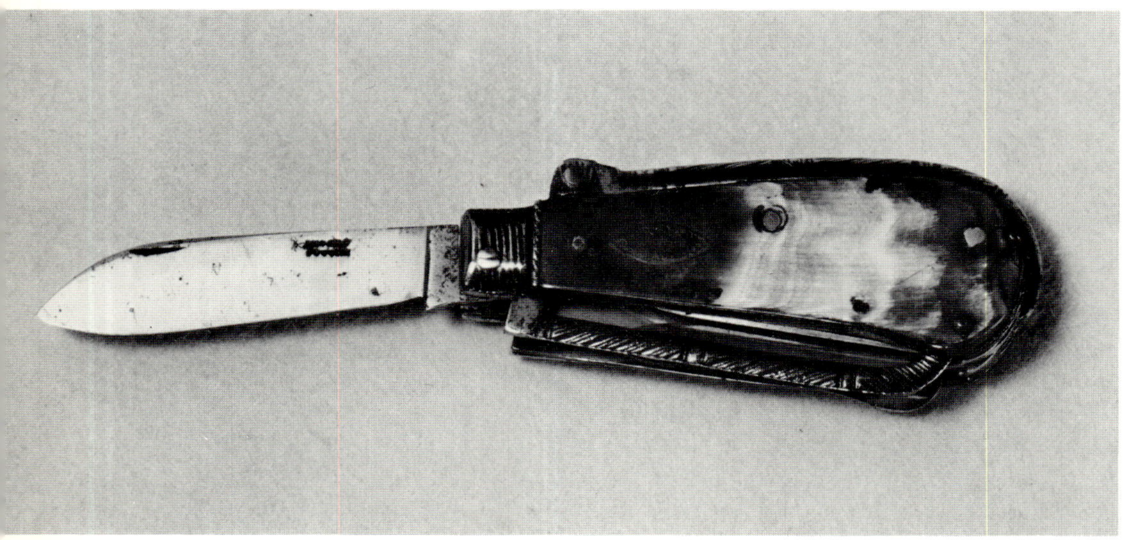

This 'multi-bladed spring knife' also has a large number of accessories. The grips are horn inlaid with silver. Joseph Rodgers & Sons, Sheffield. Late 18th century. Length, open: 5-1/2 inches. City Museum, Sheffield

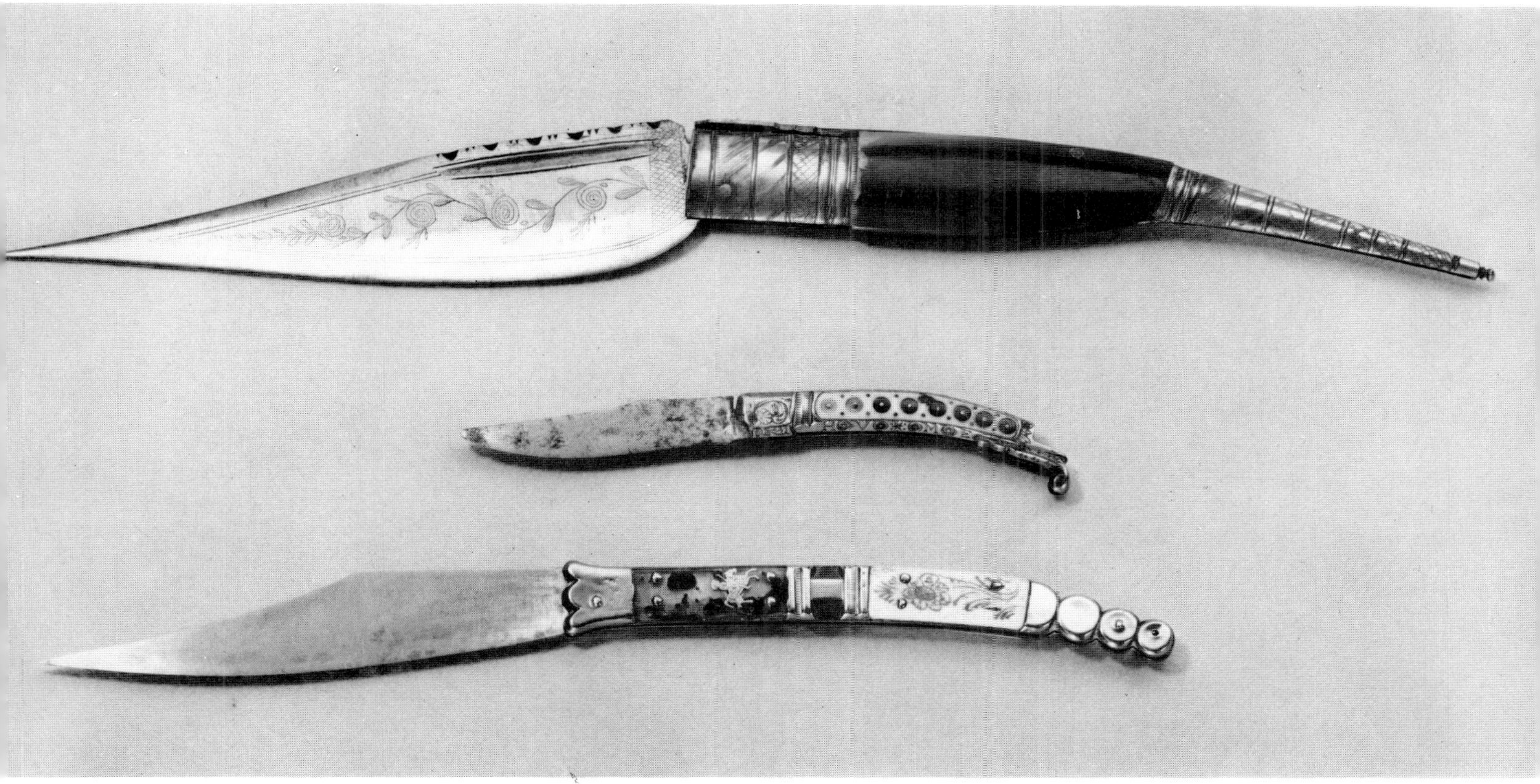

These spring folding knives are of Spanish origin. The first, with a horn and brass handle, dates from 1860. The total length: 19 inches. The second, of the same material, is older by a century. (Note the rivets and the rosettes). 8 inches. The last, with a handle of brass, horn and bone, is decorated with an inlay of a gentleman on horseback. Middle of the 19th century. 13-1/2 inches. City Museum, Sheffield

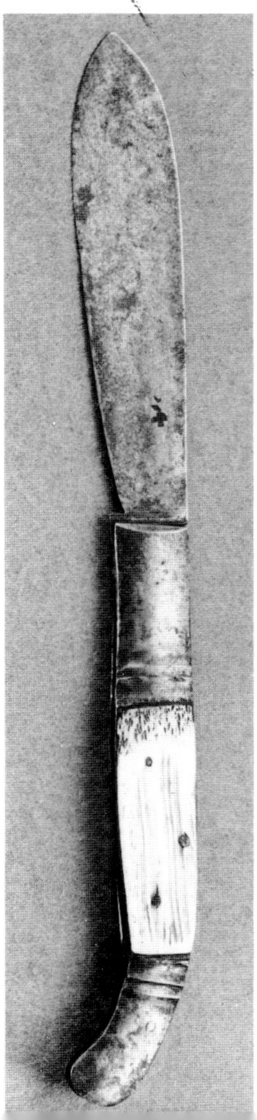

The knife on the left is of English make of the 'Barlow' type. It was widely used in America in the 19th century, and a prized possession of Huckleberry Finn in Mark Twain's novel. The bolster and metal butt are particularly long. The covers are made of bone. Probably from Sheffield. Late 18th century or beginning of the 19th. 10-3/4 inches. City Museum, Sheffield

This French spring knife has three blades and three accessories. The covers are engraved mother-of-pearl, with fittings of brass. Length open: 12-5/8 inches. 19th century. City Museum, Sheffield

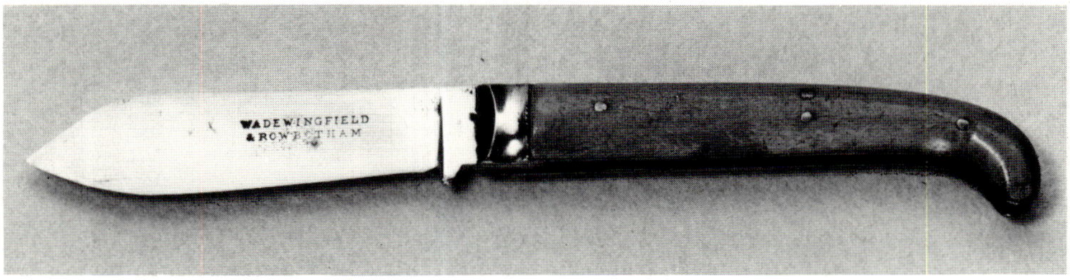

The wood handle of this English knife narrows from the heel of the blade until its end. Wade, Wingfield and Rowbotham, Sheffield, about 1840. Length open: 6-1/2 inches. City Museum, Sheffield

noted cutter writing earlier in this century: "The blade of this knife is twice as long as the handle so... when folded, it becomes a knife of ordinary size. A small hole in the blade receives a stud moved by a spring hidden in the handle and controlled by a rosette. When the handle is controlled, the blade — kept in place by the stud on the tang — becomes ten inches long, and this instrument can be used to carve meat, or if necessary, as a defence."

Thanks to the British, the picnic became a widespread practice in the 19th century. Naturally, along with it came a development in folding utensils and *outdoor folding knives*. The blade of the *pocketknife* was joined by a fork, a corkscrew, and finally a spoon, bringing a whole table service together in a small unit. All this was made possible by reducing the scale of the pieces, but above all by the precision work which appeared with the industrial revolution.

Penknives, which as their name indicates were meant for trimming goose quill pens, became portable, early in their development, and thereafter their evolution followed that of the *folding knife*. After first adding a *scraper* or 'erasing knife' and an awl for splitting the quill, another blade was added, then another, and then as many as sixteen, mounted on springs. The terms *pocketknife* and *penknife* became synonymous.

Some *folding knives* have a lock to keep the blade from closing suddenly and dangerously. Some of these are made with a powerful spring and open automatically. Such *switch-blade* or *flick knives* are dangerous weapons in the hands of criminals and delinquents. The effect of surprise makes them impossible to parry and their length and cutting edge combine to render them deadly weapons.

Knives of the *butterfly* type have a handle that separates into pivoting parts that swing around and make a case for the blade. In another system, the handle is enclosed by a wire frame; this pivots to encase the blade, and the blade, thus protected, can be pivoted into the handle.

Folding knives demand longer and more delicate fitting work than other types. Some of these could also be considered pieces of jewelry. Besides covers (or 'scales' as they are sometimes called) made of plastic, some are made of ivory, tortoiseshell, or mother-of-pearl decorated with enamel, gold, and silver. There are even some with damascened or gold-plated blades. On some knives the polished steel side covers are left bare. In the past, some covers were decorated with etchings. At the Paris World Exposition of 1900 a *Berge knife* was shown that had covers of solid gold, one blade of gold and another of Damascus steel. The handle was chased in Louis XV style. Each side had three enamel medallions made in the style of the painter Watteau. Another, which was five inches long and a little over one inch thick, had double bolsters, silver escutcheon, and 46 elements — of

The 19th century saw an increase of 'souvenir knives'. Here the handle carries busts of King William IV, Roman style, and Queen Adelaide. City Museum, Sheffield

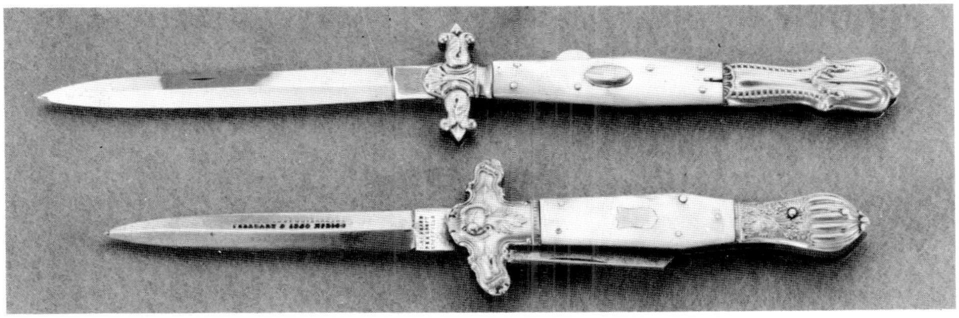

In these two folding knives, even the guard pivots. The first has a safety catch. The second is a forerunner of the switch-blade: by pressing the small blade, the large one springs open and locks. The blade of the second carries a motto. 1830-1850. City Museum, Sheffield

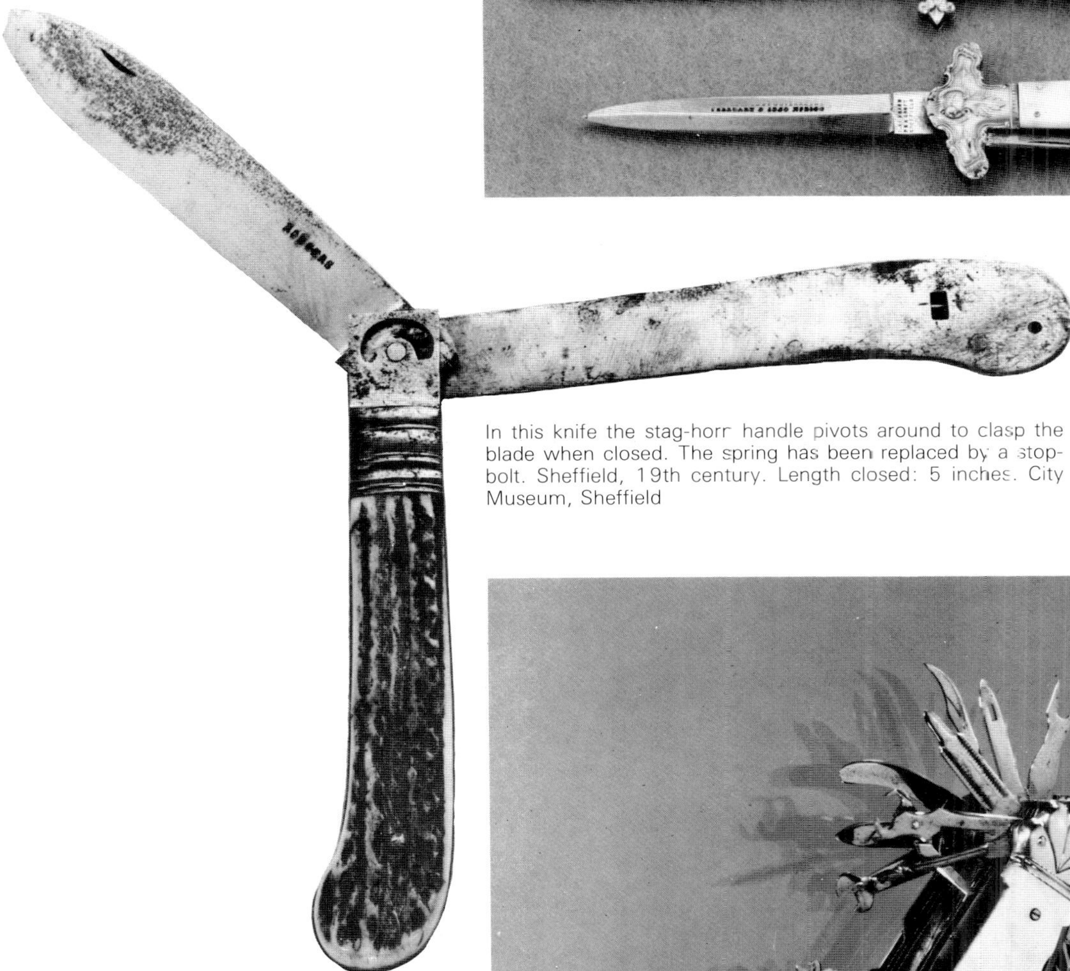

In this knife the stag-horn handle pivots around to clasp the blade when closed. The spring has been replaced by a stop-bolt. Sheffield, 19th century. Length closed: 5 inches. City Museum, Sheffield

The knife at the top comes from Czechoslovakia, where it was made before the Second World War. The second is of unknown origin. Both have covers of mother-of-pearl, and silvered bolsters. While the upper knife has fifty blade-accessories, some very unusual, the one below it has twice as many. Coutellerie du Mont-Blanc Collection, Geneva

Left: This is called a 'sportsman's folding knife', but in fact it was an exhibition model. The covers are engraved ivory. Sheffield, end of the 1870's. Length closed: 6-1/2 inches. City Museum, Sheffield

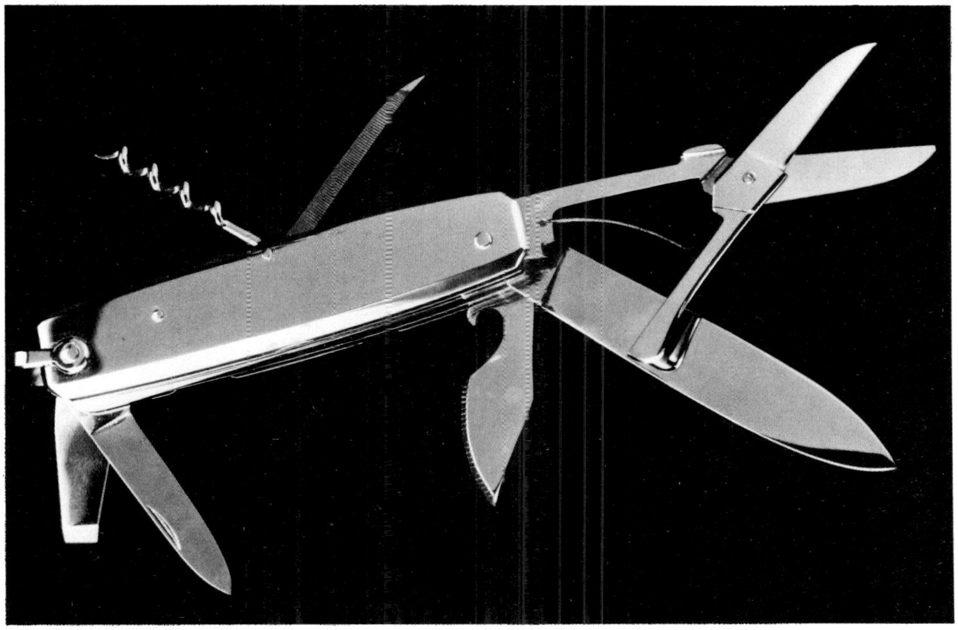

This 'Eloi Pernet knife' with eight blades and accessories is made entirely of stainless steel with a mirror finish. French-made. Coutellerie du Mont-Blanc Coll., Geneva

This folding knife is encased in grips of stag-horn. Length of the blade: 3 inches. Thickness: 1/8 inch. Dwight L. Towell, Midvale, Idaho

which 38 were spring mounted, and eight were encased in the covers. The unusual accessories included *surgeon's lancets* and a *vaccinating needle*.

This latter knife might seem rather specialized; and yet in many fields of work *folding knives* were — and still are — used as the proper tool, for example by gardeners and butchers. These kinds of knives will be covered in the last chapter, together with certain unusual models.

At the present moment the trend is toward mass production on the one hand and luxury models, or collector's pieces, on the other. In the first category, rubbing shoulders with chain-store models, are the famous 'Opinel' knives, a trademark which has become almost a household word. They come in all sizes from one inch to twelve. The patented security system consists of a bolster ring that slides over the blade to prevent it from closing unexpectedly. This knife typifies the popular knives available today.

At the other end of the scale are models which are made partly or totally by hand in the United States, Germany and France. These are mostly *sporting* or *hunting knives*. The grips are generally made of wood (sometimes plastified) or of stag-horn. Some manufacturers do not hesitate to use 'surgical steel' of the 440-C type, exotic wood, or precious stones in the manufacture. Some of these knives have parts made of brass — thus giving them a feeling of warmth and an aspect of luxury. Some are 'mixed-type' knives, that is, with some fixed blades and some folding ones.

The 'Swiss Army Knife', well-known for its sturdiness and the quality of its steel, is a descendant of a military knife made in the 1880's. Classified by price, it is in the middle range. Like other *folding knives*, it comes with models that have many pieces ranging from a magnifying glass to openers or motorist's tools.

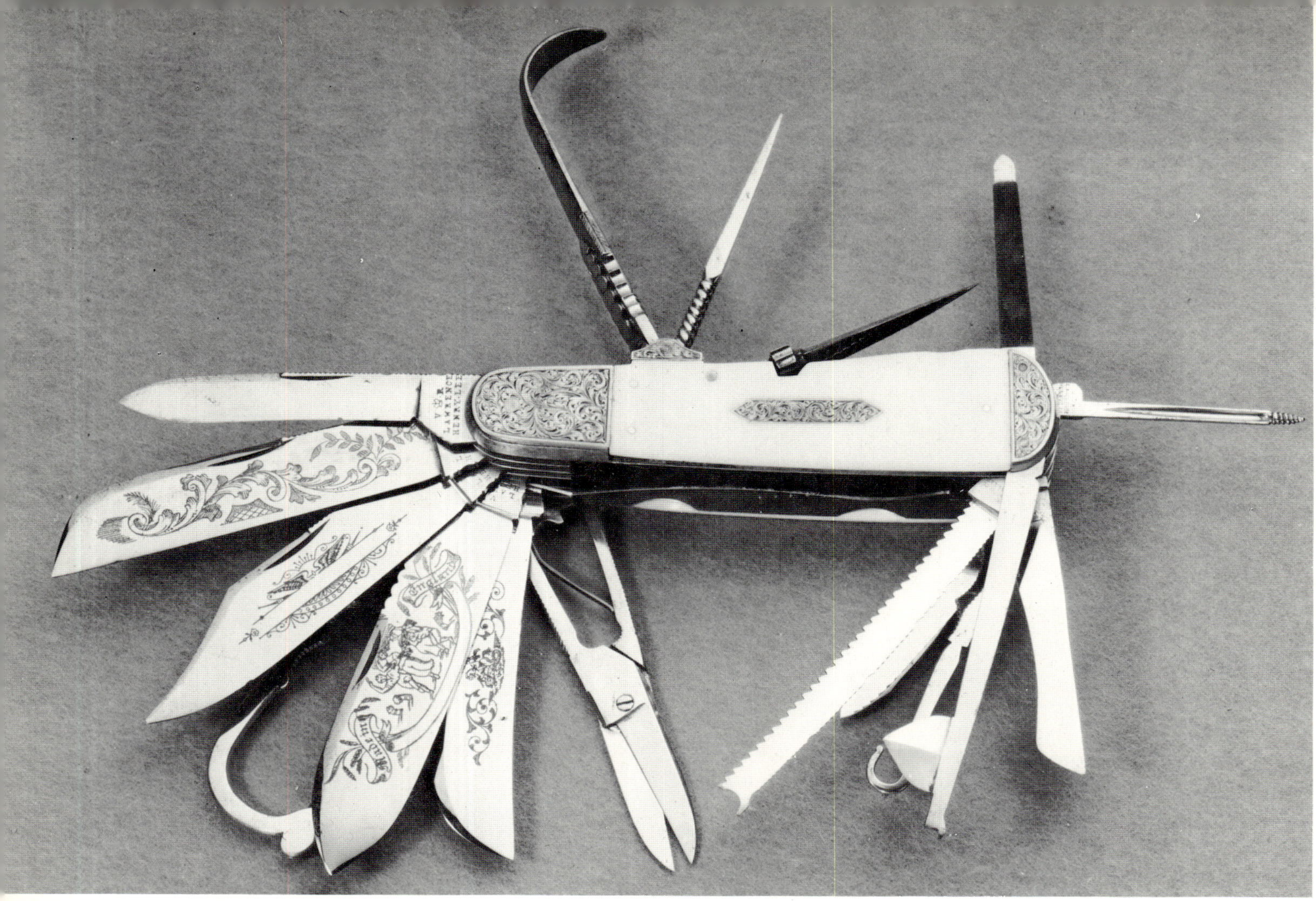

This exhibition model knife is equipped with 18 blades and accessories. The covers are made of ivory and silver. Note the curious arrangement of the blades on top. Sheffield, end of the 19th century. 6 inches. City Museum, Sheffield

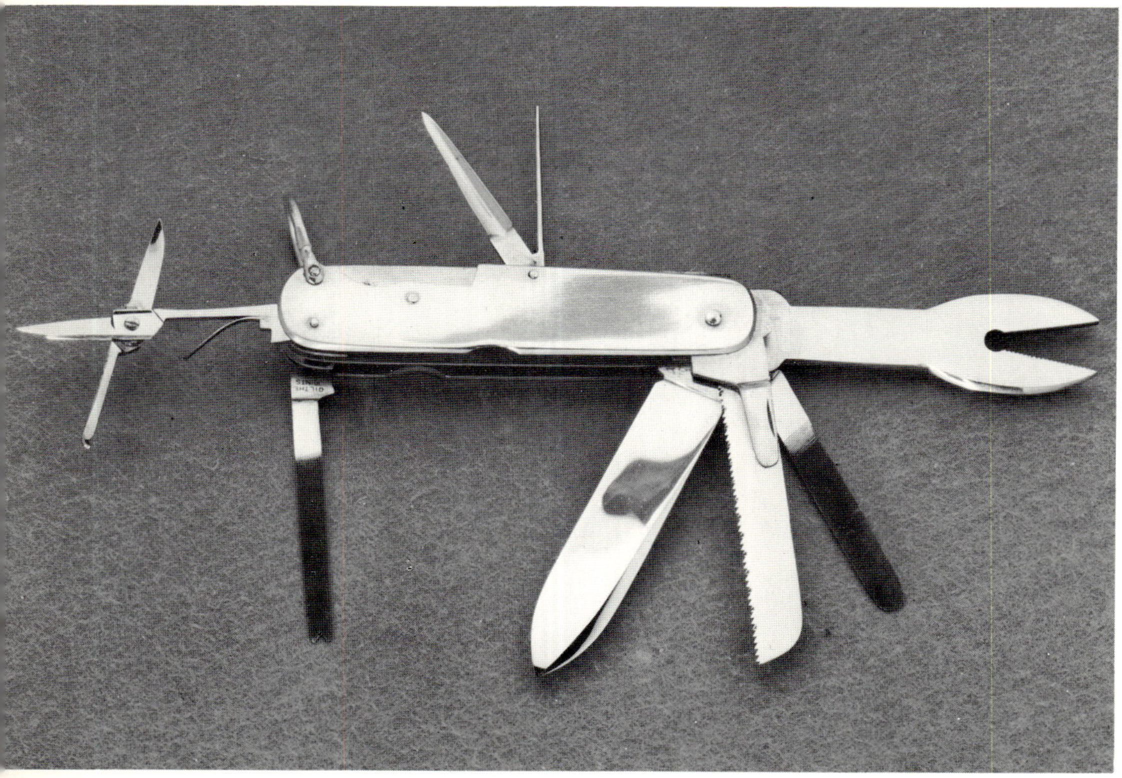

The advent of the automobile caused additional accessories to sprout on folding knives. This 'driver's' model has, among other things, a variable spanner, a feeler gauge, and a pin for cleaning windshield wiper spray jets. The covers are steel. Length closed: 4-3/4 inches. England, about 1920. City Museum, Sheffield

Heinrich Frank, an American craftsman knifemaker, specializes in folding knives for collectors. The demand for his knives is so keen that there is a ten year waiting period for each new order! He refuses to use glue for assembly and uses rivets exclusively. A good number of his knives are 'American Barlows', on which the bolster takes up a third of the handle. But it is above all the engravings that have made him famous; they have been compared to those done on firearms. The bolsters and fittings are nickel-silver and the escutcheon plates are silver. All of these knives, with one exception, have safety catches. The handles are Micarta, moose, stag-horn and wood. Heinrich H. Frank, Whitefish, Montana

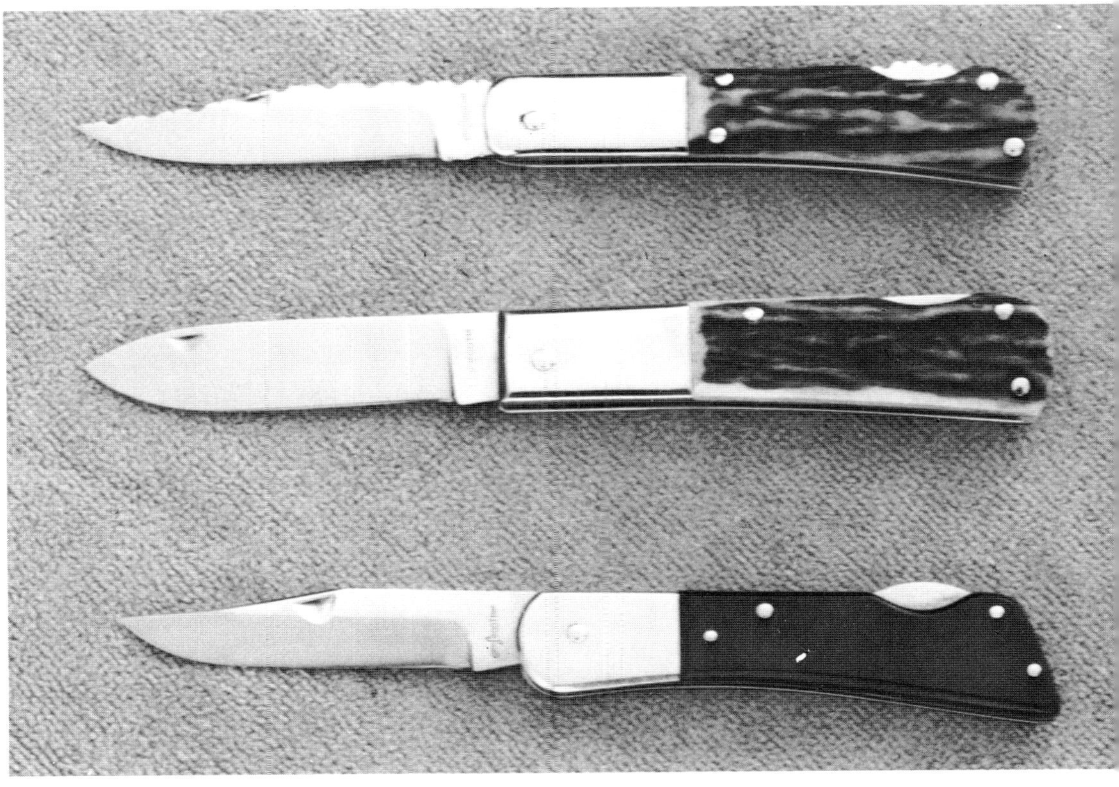

These three knives by Lew Booth form a strong contrast with the preceding selection. Booth has a special nail-nick, and prefers undecorated bolsters. In the model at the top, even the safety catch has been decorated with scallops. The handles are stag-horn and wood. Lew Booth, Boonton, New Jersey

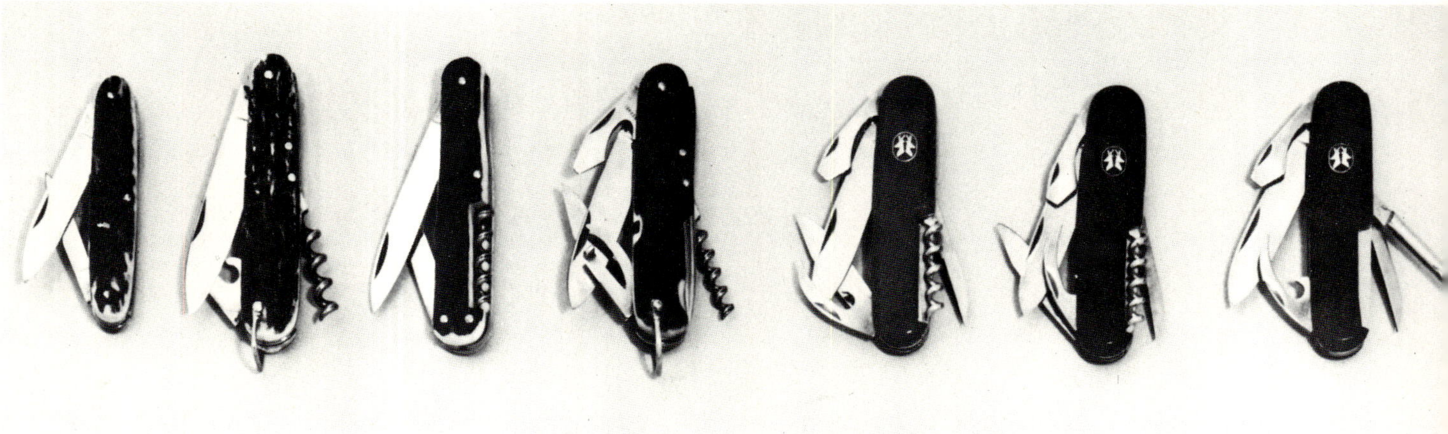

These folding knives are of a classic type. They differ in the materials used (the first four are stag-horn, the last three are plastic), the number and kinds of accessories, and their lengths (2-3/8; 3; 2-1/2 inches for all the others). Wingen, Solingen

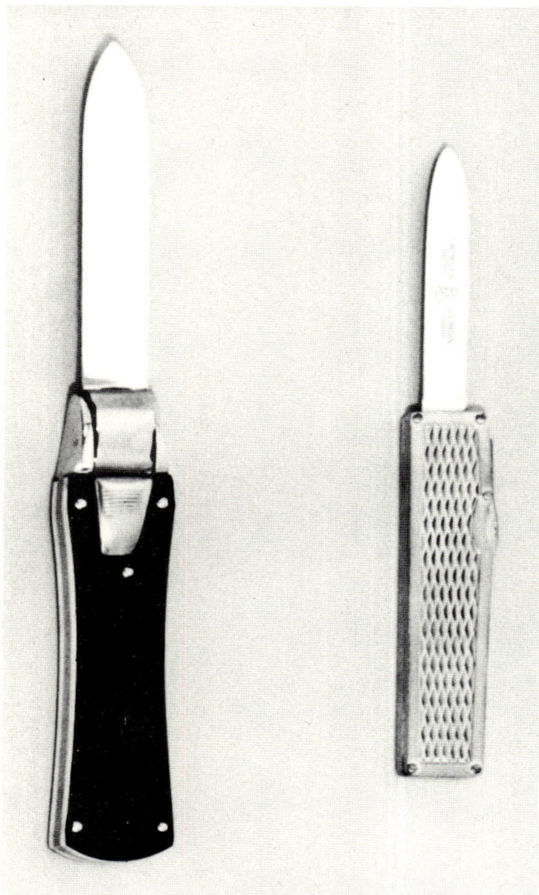

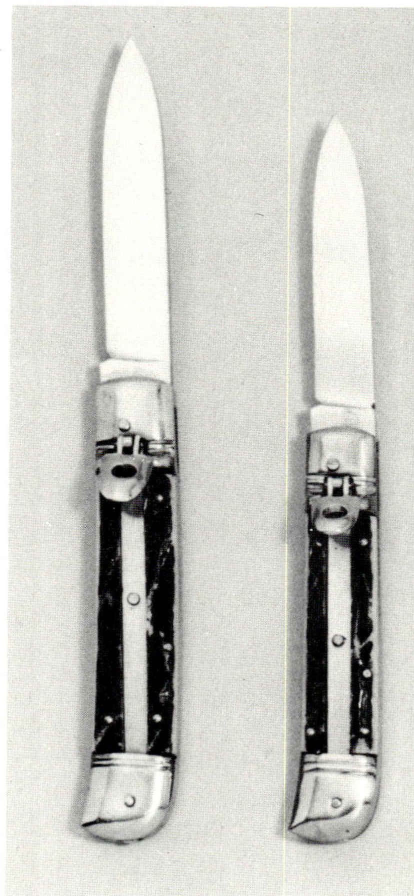

These four knives are switch-blades. In the two at the left, the blade springs out lengthwise; the knife is held vertically to close it. In the two on the right the blade springs out laterally and automatically. The locking system can be seen below the bolster. The handles are plastic, metal and stag-horn. Lengths of the blades: 3-1/4; 2-1/2; 3-1/8; 3 inches. Wingen, Solingen

A single glance at this 'model 101' reveals its sturdiness. Except for the covers, which are wood, the knife is made of stainless steel. The rivets are notched and the handle is screwed on, which is rare. Fred L. Smith, Fair Oaks, California

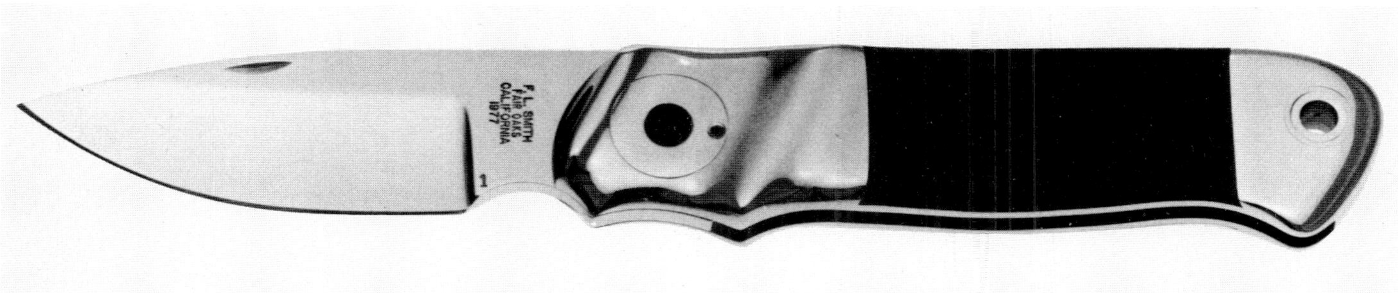

Except for the shape of the handle, and the way it is attached, this model resembles the knife on the facing page. It is also made by Fred L. Smith, Fair Oaks, California

The series of holes in this handle makes the knife lighter, even though it is all steel. It is an extra-thin model that can be entirely taken apart thanks to the screw which can be seen next to the pivot. All of Smith's models are designed to lock automatically, and protect the blade, when closed, by preventing it from coming in contact with any surface. Length of the blade: 2-3/4 inches. Fred L. Smith, Fair Oaks, California

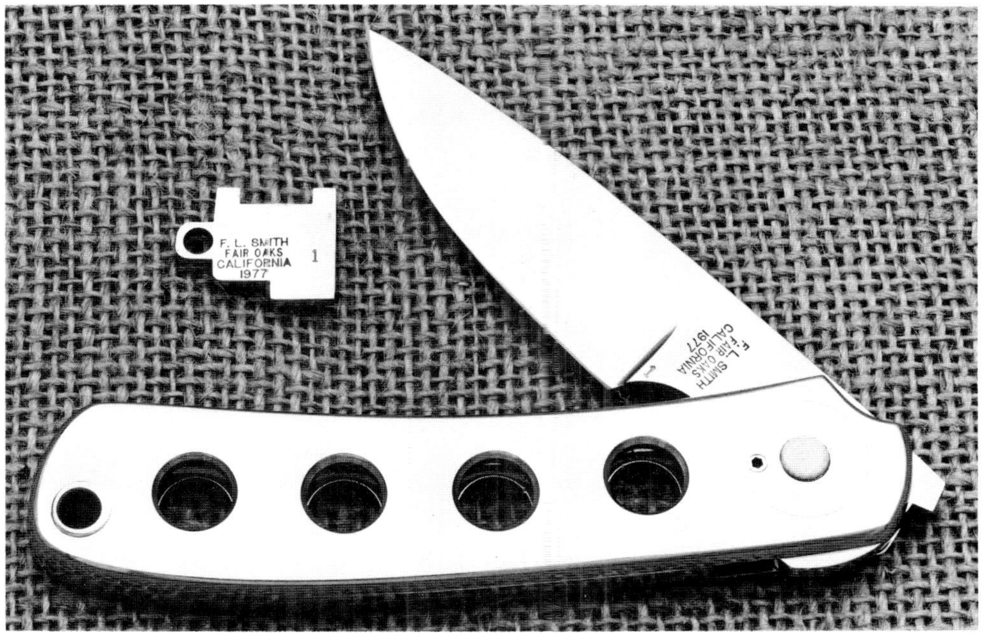

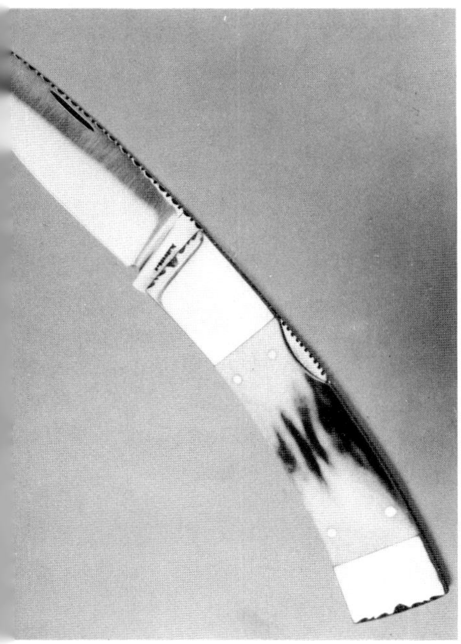

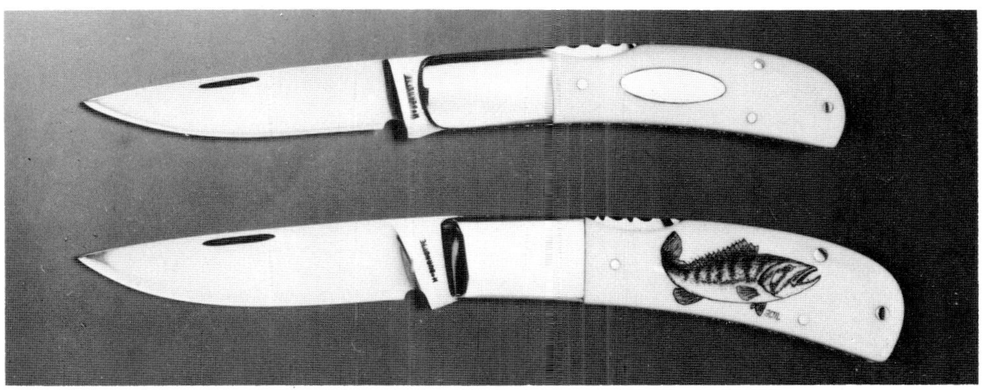

Above: This is a hunting knife with a squared-off butt. The safety catch is placed exceptionally on the forward part of the knife. The blade is D-2 steel. Thickness of the blade: 1/8 inch, length: about 3-1/2 inches. Harvey McBurnette, Piedmont, Alabama

Above the diagram: Even the decoration of these knives is the work of the maker, McBurnette. The oval escutcheon plate is meant for the initials of the owner. The blade is about 2-6/8 inches long. Harvey McBurnette, Piedmont, Alabama

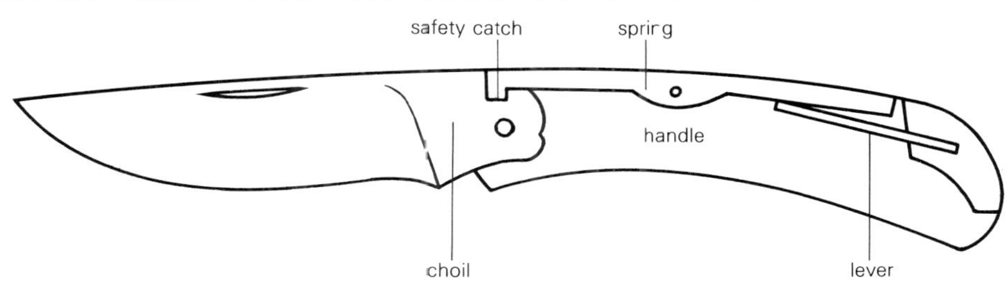

FOLDING KNIFE WITH SAFETY CATCH

Smallness of scale is not a problem for Frank: he can make miniatures, such as the one in the middle (blade: 5/8 inch) — a miniature model of the knife above it — and larger knives like the other two which have blades between 3-3/4 and 2-3/4 inches long. The engravings are masterpieces in all examples. Nothing is missing from the small model, neither the bighorn sheep nor the smallest arabesque, and the safety catch works perfectly. Frank had to make special tools in order to construct this marvel. Its handle is mother-of-pearl; the bolster and trimmings are 14 carat gold. The other three handles are elk horn, ebony and solid silver. Heinrich H. Frank, Whitefish, Montana

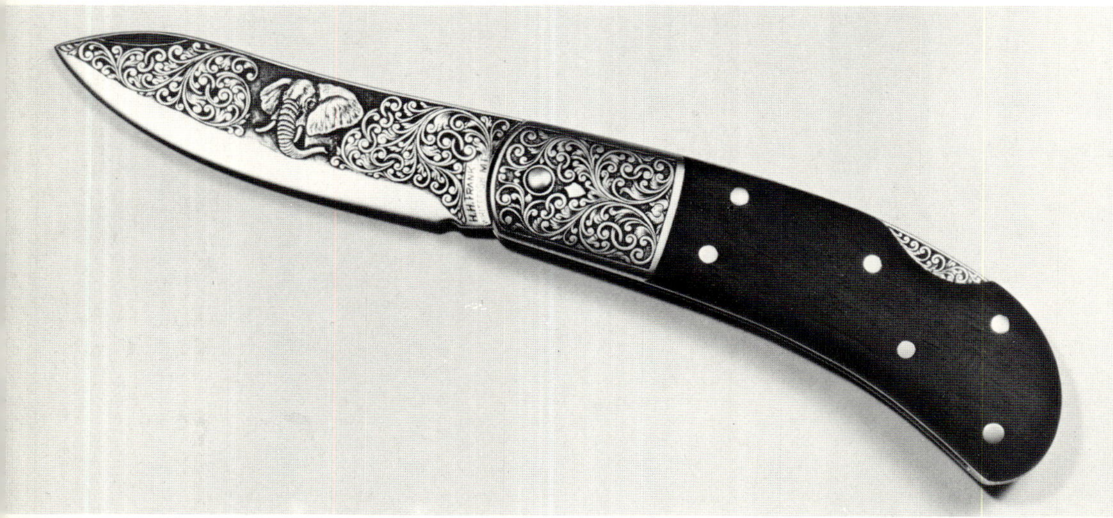

The model below, the one on the right and those on the facing page are the work of McBurnette: the unity of style, including the notched spines, stands out. Harvey McBurnette, Piedmont, Alabama

It is chiefly through their engravings that the pieces of McBurnette catch the eye (on the left a 'puma', on the right a 'five-pronger'). Harvey McBurnette, Piedmont, Alabama

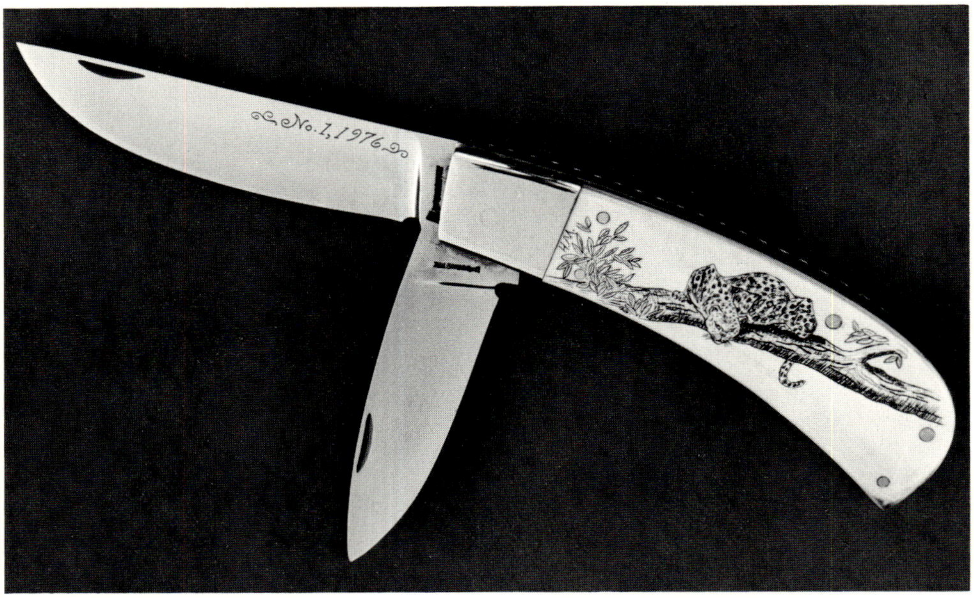

Left: This two-bladed knife is decorated with a leopard. It does not have a safety catch. The grips are ivory, the trimmings nickel-silver. Length of the blades: 2-1/2 and 3-1/4 inches. Harvey McBurnette, Piedmont, Alabama

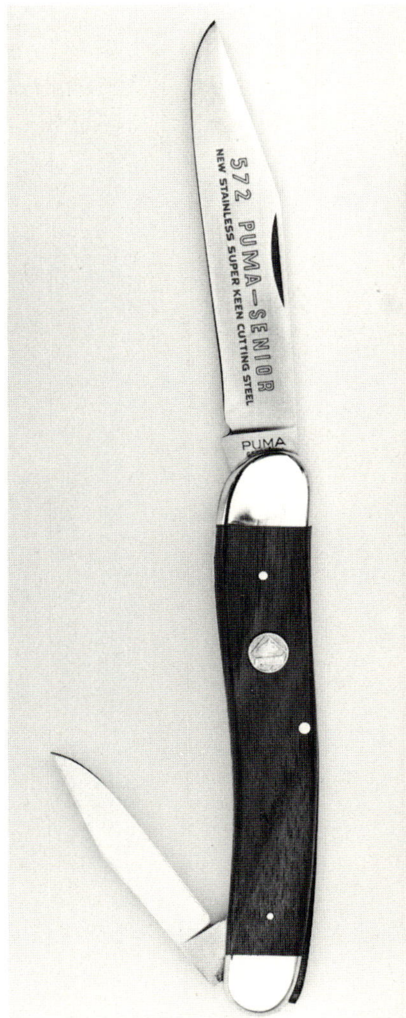

Above: The 'Puma-Senior' is truly an all-purpose outdoor knife. The handle is Brazilian jacaranda (not to be confused with rosewood) and nickel-silver. Lengths of the blades: 3-1/8 and 2 inches. Length closed: 4-1/2 inches. Puma, Solingen

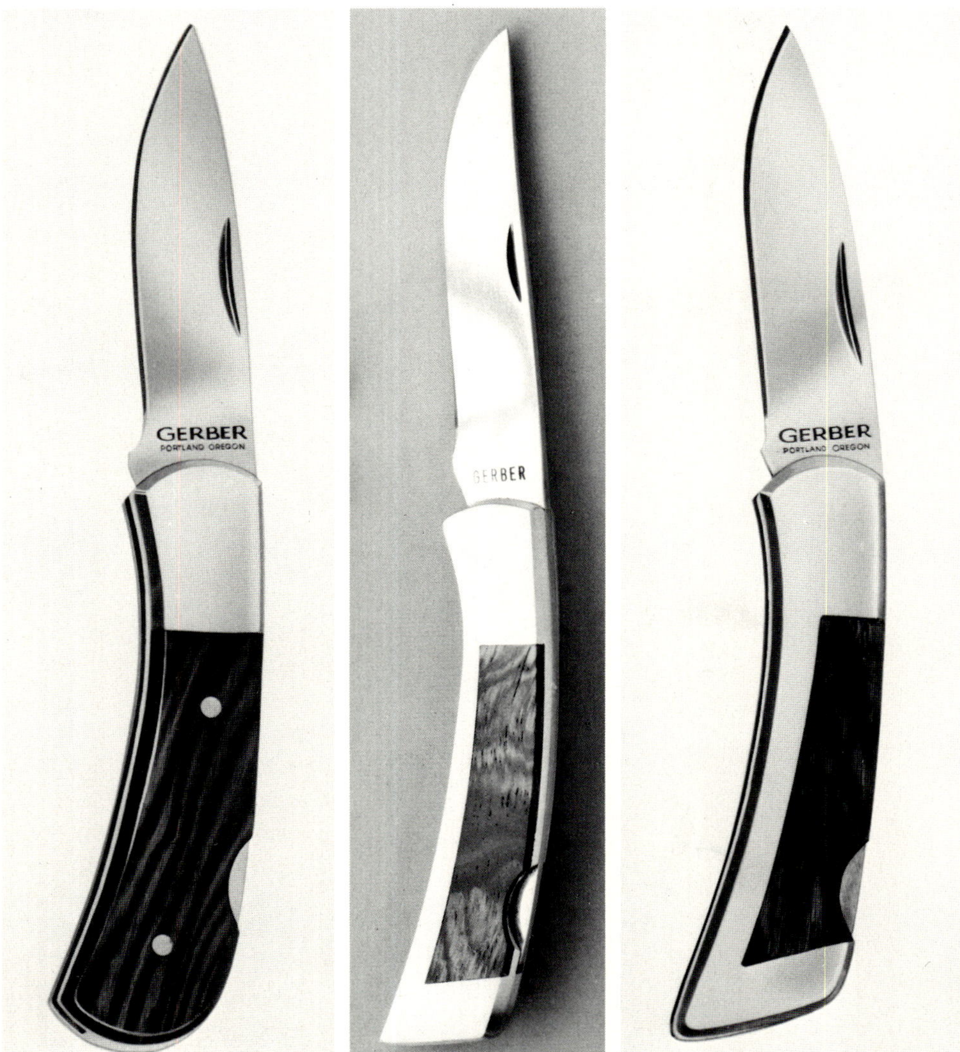

Left: These three knives with safety catches are examples of a classic shape found on both sides of the Atlantic. The handles, made of surface-treated tropical wood, are curved and grow larger in the direction of the butt. The 'Classic' and the 'Folding Sportsman I' (center) measure 6-1/8 inches; the 'Folding Sportsman II': 8 inches. Gerber, Portland, Oregon

These two spring folding knives are made entirely of steel. The covers of the one below are chased with repoussé work. Length closed: 4 inches; open: 7 inches. Billy Mace Imel, New Castle, Indiana

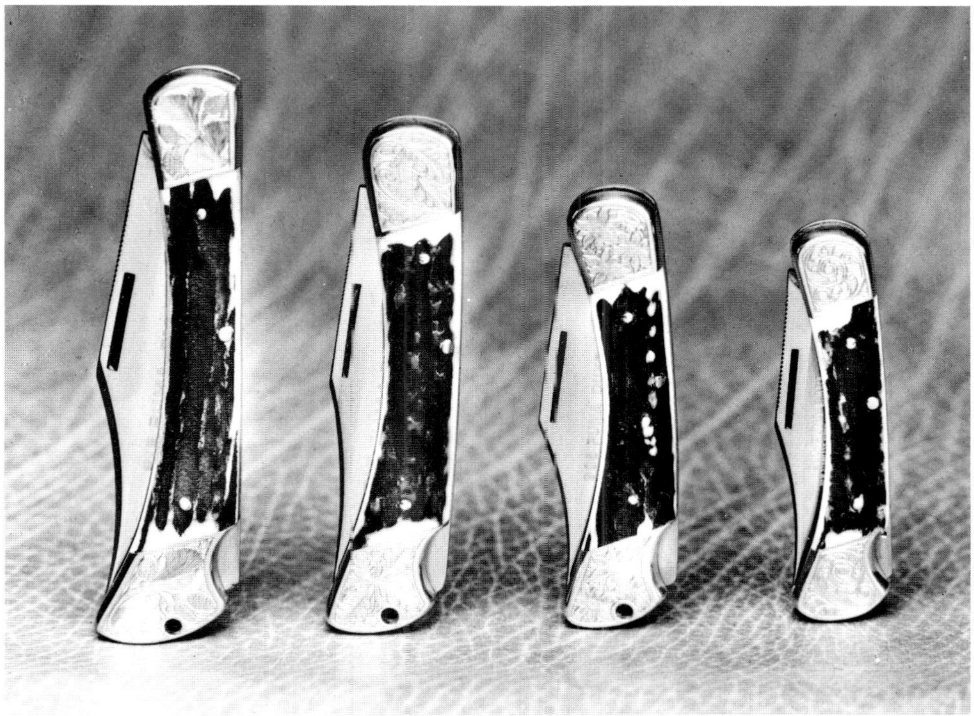

The grips of these hunting knives — all with safety catches — are stag-horn. The bolsters are brass and hand engraved. It is the marriage of these two materials that gives these knives their beauty. Note the rectangular nail-nicks and the ridged thumb-rests. Lengths of the blades: 4-1/2; 3-3/4; 3-1/8; 2-3/4 inches. Puma, Solingen

Jules Mongin, of Nogent-en-Bassigny, is known as 'the best craftsman of France', and rightly so; the beauty, polish and precision of the work in this hunting knife proves it. The knife has a safety catch and ring. The handle is horn. Author's collection

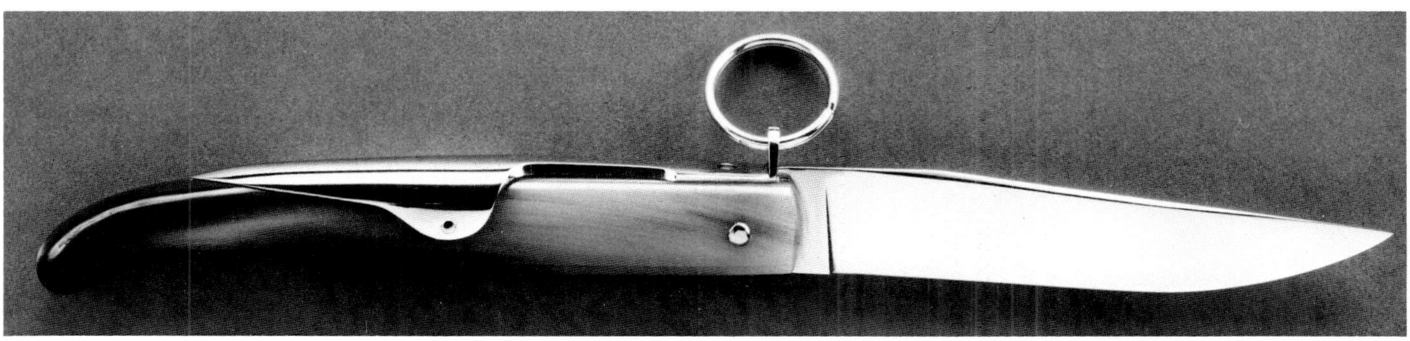

This model, intended for fishermen, is a 'weighing-knife'. When the fish is hung from the blade and the suspended knife is horizontal, the weight can be read on the blade. The blade is ideal for gutting fish. Length closed: 6 inches. Puma, Solingen

Below: This fisherman's knife can be used as a bait needle, disgorger, scaler, bottle opener, etc. Length of the blade closed: 4-3/4 inches. Puma, Solingen

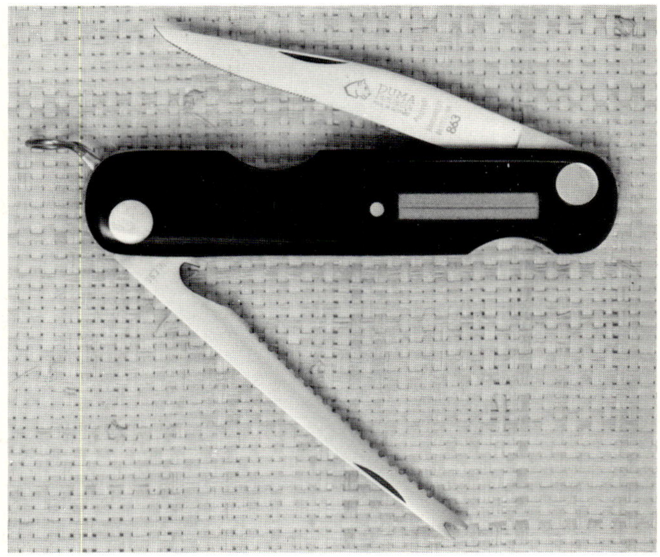

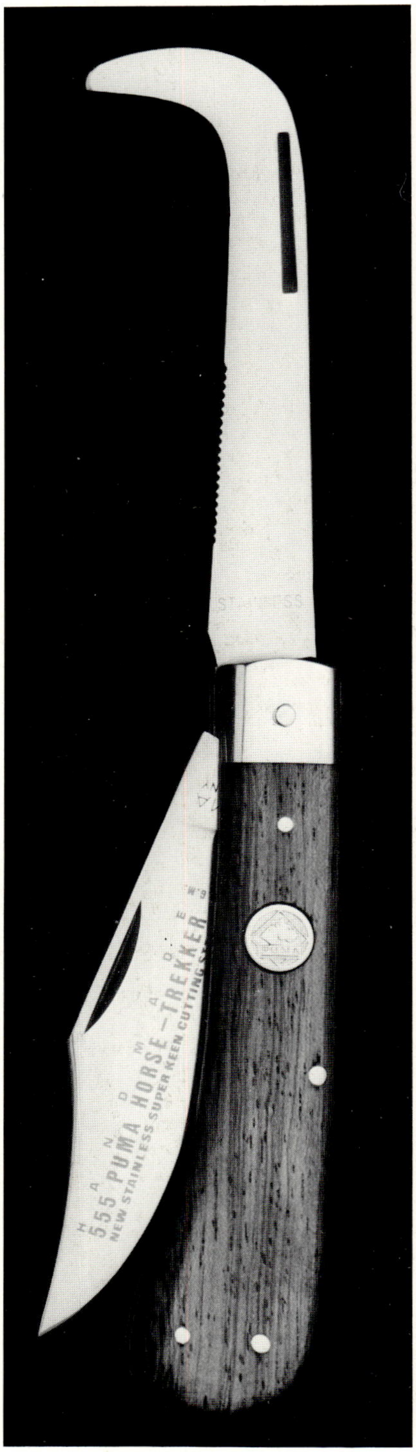

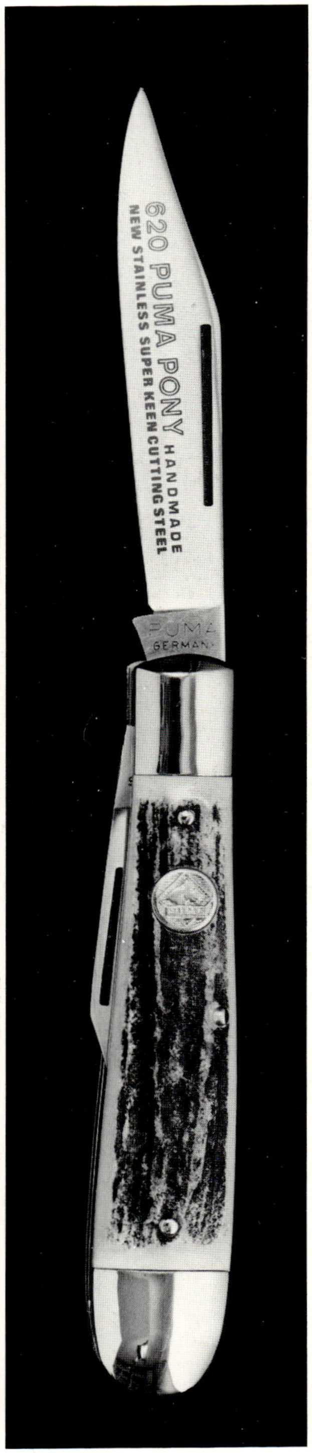

The knife on the left is the ideal instrument for the horse rider or groom. It is equipped with a hoof-pick. The handle is sealed jacaranda and the trimmings are nickel-silver. Puma, Solingen

The 'Puma Pony' is an outdoor knife with a stag-horn handle. The way it is assembled and the nickel-silver trimmings make it a very sturdy model. Length closed: 3-1/2 inches. Length of the blades: 2-3/4 and 2 inches. Puma, Solingen

Clockwise, from top: The 'Corsican Vendetta' is more a tourist souvenir than a quality knife, but it reproduces faithfully the traditional product. It does not have a spring but has a very long bolster; the handle is of synthetic material and hand painted. The knife with a red plastic handle is Swedish and has a safety catch. In the following American model by Case, the guard pivots with the blade to fold away with it. The knife with a large and impressive blade and a horn handle is a Spanish model with a safety catch. The red handle of the Finnish hunting knife splits into two parts to close around the blade. At the bottom is an Italian switch-blade. It has a safety catch to keep the blade open, and a lock to keep it closed. One of the buttons makes the blade spring out automatically and laterally. The second Spanish knife has a shape which is more classic than the first. Again, the handle is horn. The red knife with its 'piano wire' style handle is a turn of the century American scout model. Finally, the knife with the rustic style blade is a hand-made Chinese knife. The ring acts as a safety catch. The rounded part is the back of the blade, the straight part the edge. This knife does not have a spring. Coutellerie du Mont-Blanc Collection, Geneva

Vendetta Corsa

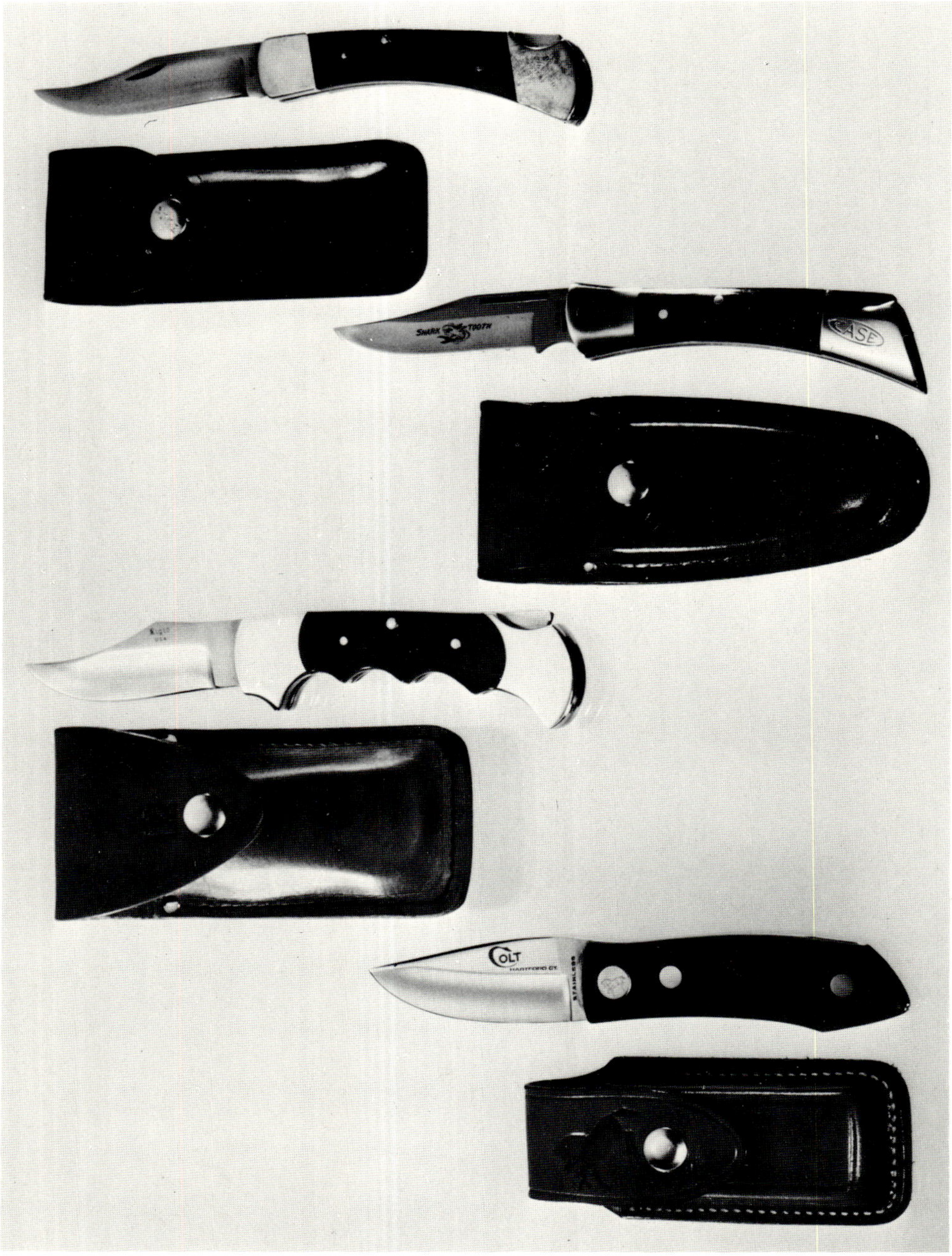

Left: The first three knives have safety catches. The handles are rosewood and wood conglomerate. The trimmings are brass. The first is a 'Buck 110', the second, the famous Case 'sharktooth'. The third is the Rigid 'Apache'. Note the difference in the shapes of the nail-nicks and the handles — flat or rounded, with finger grips or without. The last one by Colt is a 'butterfly' knife. The two grips or covers pivot to encase the blade. American made, Coutellerie du Mont-Blanc Collection, Geneva

Right: The knife above is a marvel, with a handle of first class horn and the steel mirror-finished. It is 6-piece French knife by Mongin. The back is steel. The three others are Swiss knives. The popular 6-piece model on the left is based on the old military knife on the right. These Victorinox brand knives are known for their sturdiness and quality of blades. Coutellerie du Mont-Blanc Collection, Geneva

The 'Opinel' is one of the most popular of all knives. It comes in all sizes from about 6/8 inch to 12 inches. The smallest ones have plastic handles, the others are varnished wood. Some blades are stainless, others, carbon steel. Most have the famous sliding safety bolster.

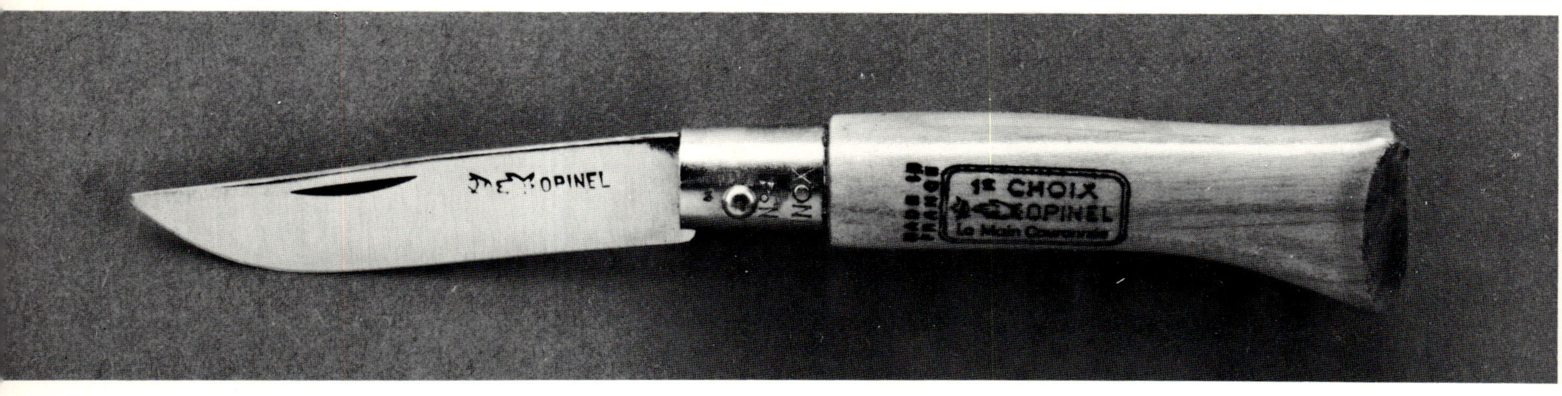

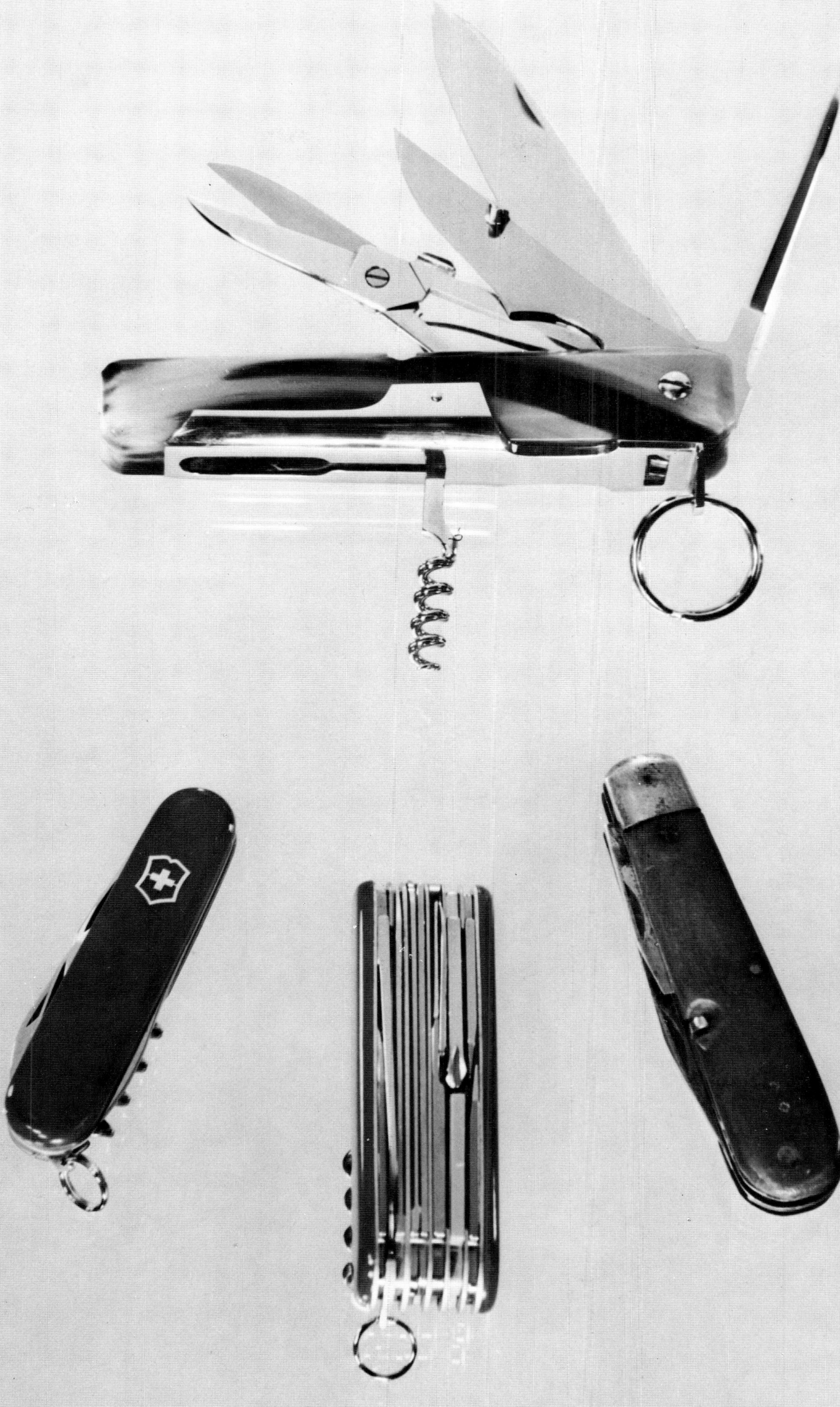

SPECIAL, UNUSUAL AND EXOTIC KNIVES

"Imagine to yourself a large hall filled with guns and sabres from top to bottom; weapons of the whole world: carbines, rifles, blunderbusses, knives..."
(Alphonse Daudet, *Tartarin de Tarascon*)

Butchers' knives

Besides *table* and *kitchen knives*, cutlers make a number of special cutting instruments, some of which are quite far removed from the usual.

To begin with, the *butchers' knives* may be considered. Butchering implies stock keeping, which replaced hunting. *Butchers' knives* are intended for two principal functions which are today quite distinct from each other: slaughtering and butchering. The methods of slaughter vary according to the country, the animal in question, and the size of the slaughterhouse. With the exception of Jews and Moslems, who bleed the animals to death without stunning them first, the slaughter of large livestock implies the use of the pole axe, cartridge activated stunner, or captive bolt pistol. In some cases, a sticker may be used to sever the base of the brain from the spine. At some point in the process the animals have their throats cut and are bled using a pointed, long and narrow knife.

The *flaying knife* is relatively small with a wide blade. The knife meant specifically for pigs has a rounded end. For sheep there is also a special knife and even a special vocabulary for some of the operations, which are distinct from those used for cattle. The *sheep knife* has a short cutting edge which forms a quarter of a circle at the end. The *scaling knife* with a square blade is used for scraping the hides of pigs. It has a rounded edge to help clean around the joints while keeping the hide intact. Many variations of each specific kind of knife are possible.

For actual butchering there are many types of knives, but they generally have wide blades that narrow towards the point. Exceptions are the *boning knives*, which have to be narrow in order to fit between the joints, as do the *veining knives*. There are also whole series of very specialized butchers' knives such as *sausage knives* with short blades, and the very wide *bacon knives*, used for preparing ham and sausages.

The Jules Dassin film 'Topkapi' made this knife the symbol of the famous museum of Istanbul. Here the knife has become a piece of jewelry: one of its sides is ornamented with three emeralds 1-1/4 to 1-1/2 inches long surrounded by diamonds. The scabbard is gold decorated with diamonds and enamelled flowers. The pommel of the handle is capped with an octagonal emerald which covers a small watch. This piece was to have been a gift to Nadif Shah, Emperor or Persia, but the gift never arrived at its destination. Total length 14 inches. Ottoman Empire, 18th century. Topkapi Museum, Istanbul

Bakers' knives

Before leaving the world of food we can mention the *baker's knife*, which is very thin and sometimes lacks a handle. It is used for tracing light lines on bread to open the dough and let gases escape during baking. The *pastry cook's knife* comes in several shapes. Usually the end is round and the two edges of the blade are parallel. Sometimes the cutting edge has fine teeth. In this category are also *tart slicers*, sometimes sharp-edged, and *spatulas*.

Surgical instruments

It is perhaps natural to discuss surgeons' instruments just after butchers' and bakers' tools, since for centuries surgeons were considered tradesmen because they worked with their hands. Originally they were in the same guild with the barbers. In the time of the surgeon Ambroise Paré (1509-1589), the doctors were clerks, but under the pretext that '*Ecclesia abhorret a sanguine*' (the Church abhors the shedding of blood) they would consider it improper to let the blood of a patient. The surgeons were not fully free of this restriction until 1743, after which they were considered full doctors.

Oribasis (fourth century), a famous Greek surgeon, knew how to trepan skulls — and the instrument with which this was done could well be more ancient than the *circumcising knife*. He also used the *scalpel*, various other surgical knives, tools, saws and the cautery. Such treatises as those of Abul Kasim (c. 936- c. 1013) of Cordova and Jean Scultet (1595-1645) may be passed over: the description of their arsenals bring to mind instruments of torture rather than healing; but some of the instruments they mention exist, barely changed, up to today.

An excellent work, which complements that by the 18th century cutler Perret, is *A Short Description of Surgical Instruments* by 'Henry, Cutler to the (French) House of Peers' (1825). The author describes 200 instruments and apparatuses, classified by type of operation. For childbirth he enumerates forceps, hooks, lever, head piercer, head clamp, the *speculum uteri*, the *scalpel*, the *long blunt-ended knife*, the pelvimeter and laryngian tube. The *straight knife* is used for sectioning the symphysis of the pubic bones; the knife must be strong, fixed without the slightest play on its handle and blunted at its end. As for the *scalpel* used for vaginal caesarean sections, it must be 'rounded and wrapped with stripping up to an inch from its end'. In abdominal caesarean sections 'a *convex* and a *blunted scalpel* are used'.

In the beginning of the 19th century various knives were used for amputation. In the amputation with flaps, more bone is cut than flesh in order to obtain a presentable and sometimes usable stump. The first knife used in such an operation is slightly curved. Later a *simple-edged straight knife* was used, with a slight curve on the spine towards the point. Its edge is similar to that of a *scalpel*. For this knife, the surgeon Garengeot substituted one with a straight blade with two edges. The surfaces were not flat, but on each side was a sharp ridge running from the heel to the point. These ridges bevelled out — giving strength to the blade and making the edge very durable. The point was thin, long and particularly sharp. In order to cut flesh and to saw with the same instrument, a knife was invented in Germany in the last century which had sawteeth on a portion of its spine, the rest of the blade being two-edged.

The 'kernelers' or nut-openers (here an unusual example in the form of a key) were in fashion from the 14th to the 17th and even 18th centuries; they were generally made of bronze and their small size was proverbial. Drawing after Victor Gay (Glossaire archéologique du Moyen-Age et de la Renaissance).

Professional knives of the 18th and 19th centuries. 1: a cheese cutter (which, strictly speaking, is not a knife); 2: a cod fisherman's knife; 3: a 'bistoury'; 4: a scalpel; 5: a preserving knife; 6: a 'cutting needle' (a surgical instrument); 7: a scalpel-razor; 8: a gardener's knife; 9: a curette for ears; 10: a lancet; 11: a multi-piece veterinary lance; 12: a 'bill-hook'; 13: a wire cutter; 14: a basket-worker's knife; 15: a painter's knife; 16: a gilder's knife; 17: a sailor's knife; 18 and 19: net maker's knives. After Jules Pagé

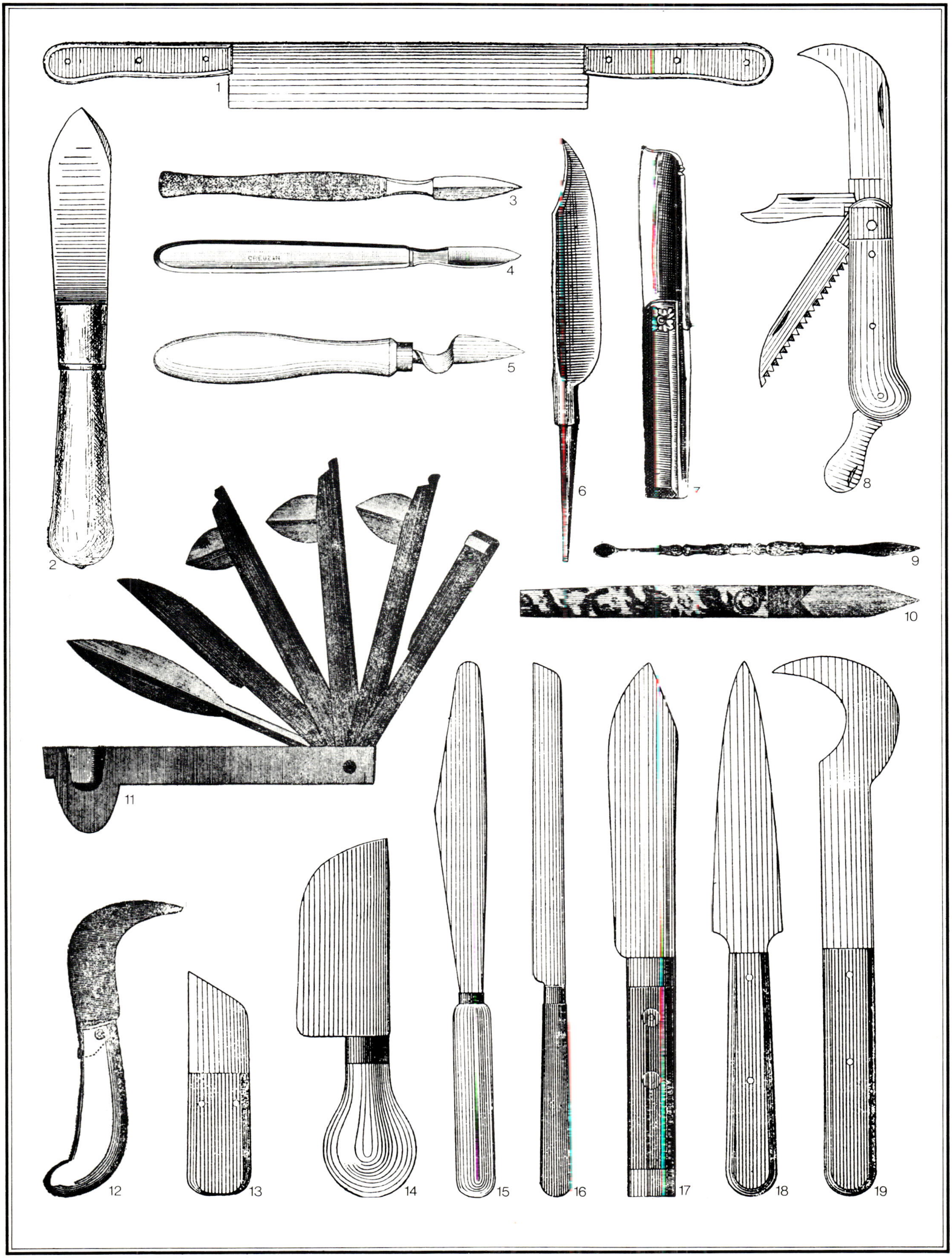
1
2
3
4
5
6
7
8
9
10
11
12
13
14
15
16
17
18
19

Surgical instruments must be of high quality, with high carbon content (more than 0.8 per cent) and hard-tempered. They must be strong and yet sharp as razors. The situation today is far different from the days when *scalpels* for use on kings were made of gold and silver. In the times when they were made by hand methods, the problem in making such precise instruments as *surgical knives* was mainly one of balance, which plays an essential role in these cases. The craftsman had to take special care to hammer them straight on and to give as many blows to the back as to the middle of the blade and to the edge.

One of the most usual forms today is the *straight scalpel* with a sharp point. For certain operations there are also *double-edged scalpels* with sharp or rounded points, as dictated by their specialized functions. The list of all of the instruments in use today, from the *scalpels* and the *lancets* to the *scarificators* and *cataract knives*, is practically endless.

Other specialized knives

There are still some other knives associated with crafts, even though growing mechanization has been eliminating these tools one by one... and the craftsmen as well. The main tool of a shoemaker is a *cobbler's paring knife*, a flat blade ending in a keen point, which is used for cutting leather. The blade is slightly concave and the knife is held firmly by the handle while the knife is pushed forcefully. Among other knives used for leather is the *skiver* or *leather shaving knife* which has a rectangular blade and is used by glovemakers to prepare thin and even pieces of leather. Bookbinders and leathergoods makers at one time used their own paring knife for 'preparing' leather so that it was even and of the right thickness; it had a large flat blade.

Below left: These surgical instruments are probably French and date from about 1800. The handles are wood and ivory. The first three are an ear probe, a scalpel, and a retracter; the last three were for eye operations. City Museum, Sheffield

Below: These three knives are manicure instruments. 19th century (?) Schloss Nymphenburg, Munich

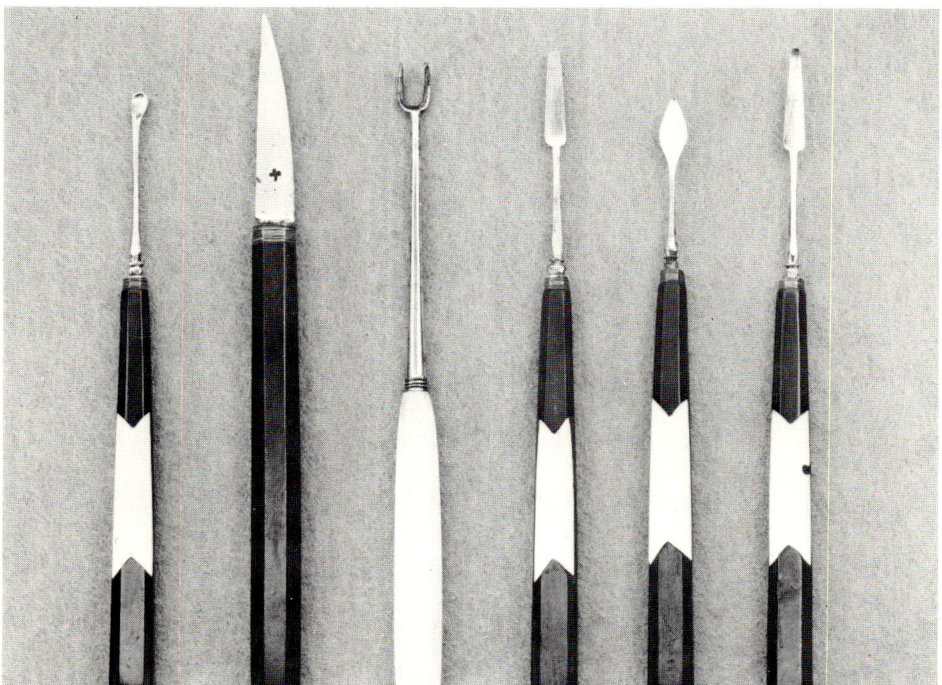

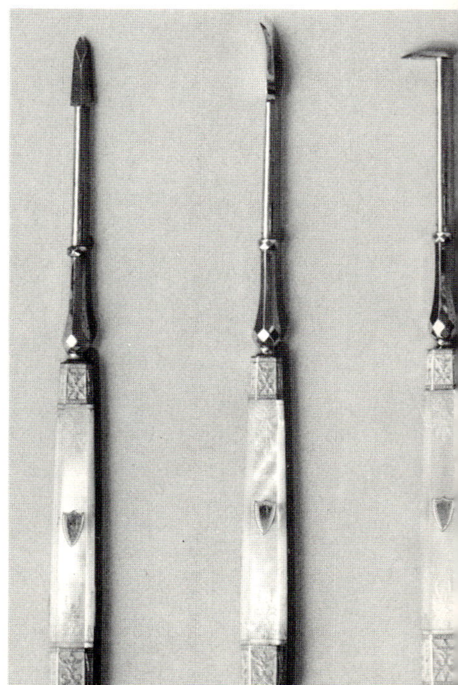

These butcher's knives show a certain diversity but many of them resemble kitchen or veterinary surgeon's knives. The first five are boning knives, the next three are skinning knives, the last are for bleeding the animal. They come in all sizes. Even knives that are meant for the same purpose can have different shapes and curvatures of the blade. Note the safety ridge on the handles. Zwilling, Solingen

Left to right: The first piece is described by the manufacturer as a grafting-knife and the others as gardener's knives, but in fact, the last three can be used for a variety of tasks. The handle of the first is synthetic, the others natural or stained wood. Lengths of the blades: 3-1/2; 2-3/4; 3-1/2; 3-1/8 inches. Herder, Solingen

The *furrier's knife* has a large blade and the heel is used to trace out the shapes of the pieces into which the leather will be cut. Still in the domain of garment-making is the *velvet knife*, which is used for making velveteen. It has a particulartly long blade, so much so that it has even been considered a specialized sword, like the instrument used by tailors for cutting cloth. In textile factories a very short *spinner's* or *threader's knife* is used to cut the thread on the spools when they begin to run out.

Professional fishermen use several kinds of knives: some are meant specifically for cod fishing and have a strong and heavy lead bolster-ring which is cast on the knife itself at the time the blade is fitted into the handle. This bolster gives great strength to the knife, and its weight makes it possible to cut off the head of the fish in one stroke. The knives of herring and sardine fishermen are very simple; the blade is merely stuck into the wooden handle, which has a hole for a loop thong.

The prime tool of the gardener is the *billhook* or *pruning hook* with a curved blade used for trimming tree branches and vines. A *grafting knife* has a small blade two or three inches long and rounded at the end on both sides. At the end of the handle is a *spatula*, called a *budding knife*, which is used to raise the bark in order to slip in the bud. What is sometimes called a *gardener's knife* is a multi-purpose folding instrument which has a knife blade, a *grafting* and *budding knife*, a *billhook* and a saw.

This model with a semi-cylindrical wooden handle is a saddler's knife. Many professional knives resemble each other. Length of the blade 4-3/4 inches. Herder, Solingen

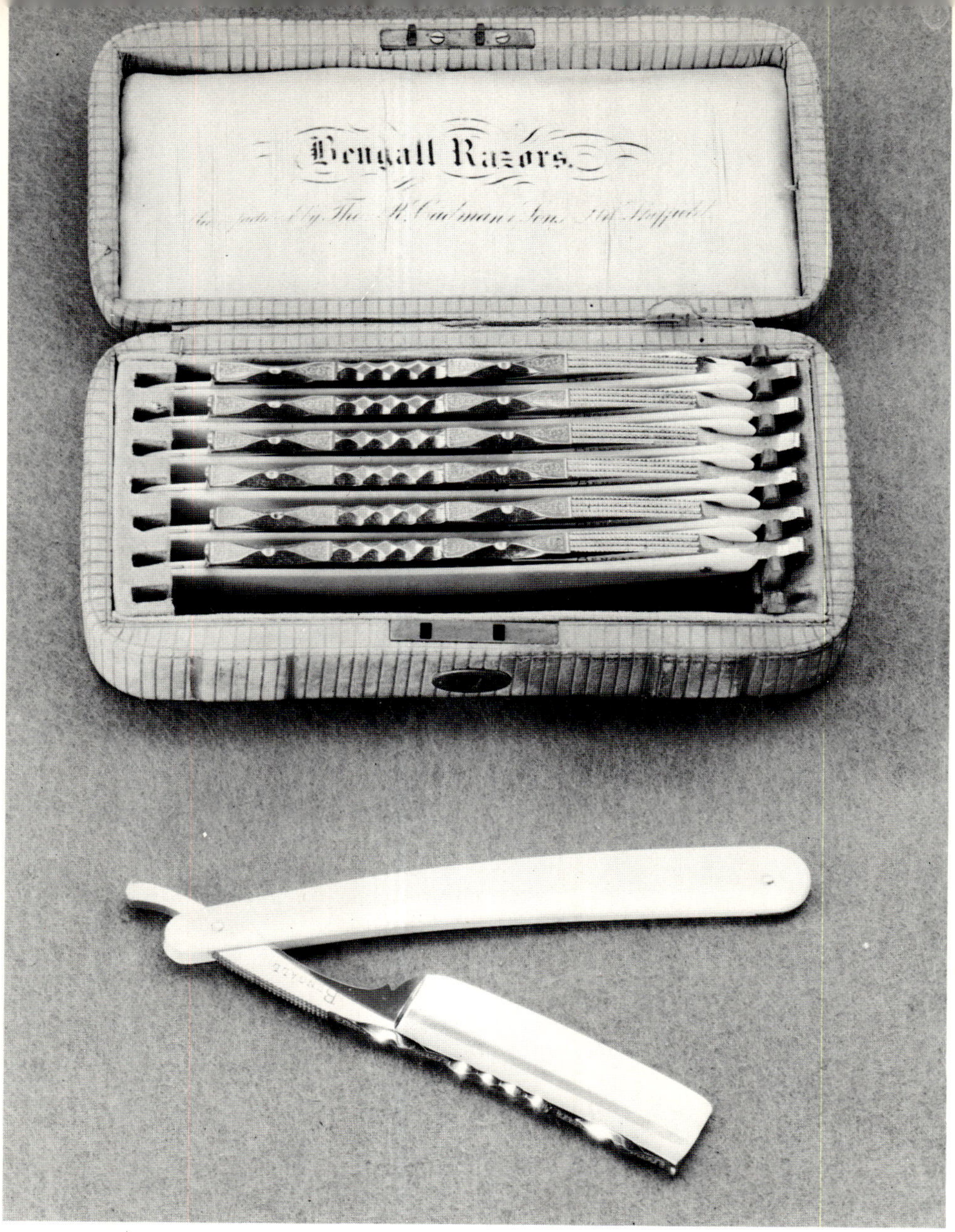

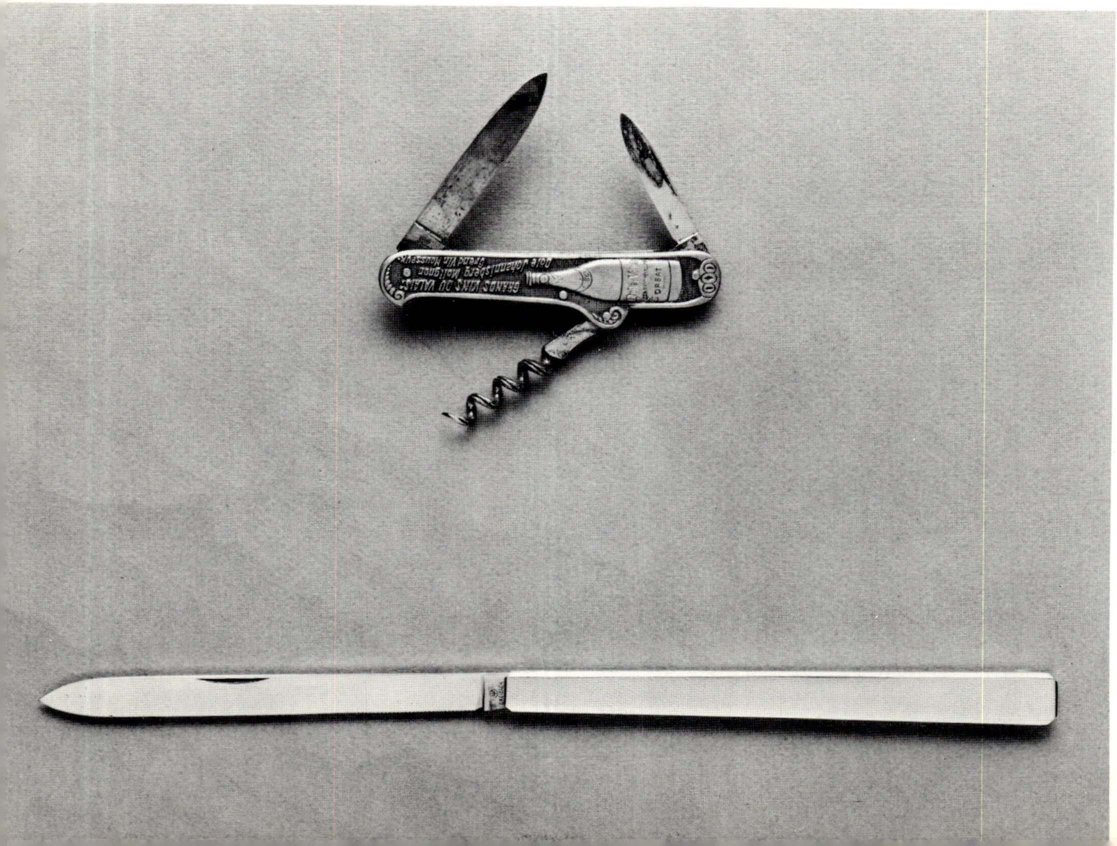

Above left: In the 19th century, the 'weekly razor set' was a fashion with the upper class. It was a case containing seven razors — or at least seven interchangeable blades — each marked with the day of the week it was to be used. The set here was made for exhibition. The handles are ivory. Sheffield, about 1880. City Museum, Sheffield

The two razors above, one in ivory and the other in mother-of-pearl, date from the beginning of the 19th century. They apparently both come from Sheffield. Length closed: 5-1/4 and 5-1/8 inches. City Museum, Sheffield

The blade of this pork butcher's and sausage maker's knife (below left) has teeth along some of its length. The handle is checkered. Length closed: 5-3/4 inches. German-made. The knife with advertising on top was meant for waiters. Length closed: 3-3/8 inches. Swiss made. Coutellerie du Mont-Blanc Collection, Geneva

Unusual and special knives

The best-known specialized blade is the razor — even though it is being rapidly replaced by its mechanical and electrical versions. As early as the 18th century the cutler Perret had created a *plane-razor* which made it simpler to shave without cutting the face. Even a *glass razor* has been invented, but all of these more or less extravagant models have been replaced by the *safety razor*. The 19th century saw the first razor with a replaceable blade and the *semainier* or a 'weekly razor set' which was a container and a set of seven razors, each engraved with the name of the day of the week it was to be used. Still in the domain of toiletry articles, there is the *powder-knife* which is nearly obsolete, and the *corn-cutter*. These resemble the blades of *penknives* but are wider and shorter. The point is rounded and curved backwards.

Paper-cutters come in all shapes: some are reminiscent of stilettos, others of daggers and even scimitars. They are made out of all kinds of materials since they do not have to be sharp. There are very small ones for cutting the leaves of new books that have a slit into which a sheet of paper can be slid. In that way they serve both as paper-cutters and bookmarks.

Most of the ancient knives which have fallen into disuse by their very antiquity or by the disappearance of the functions for which they were made appear unusual to us at first glance. Such is the *horseshoe* or *antennae knife* of the Hallstatt and la Tene cultures named after the curious appendages on either end of the handle, which made it resemble rolled-up butterfly antennae. These were for providing a better grip and easier fastening to a belt.

In the 11th century, so-called *liturgical knives* existed. They were used by the church as gifts for anniversaries and investitures.

Among *ritual knives*, one of the most ancient is the *circumcision knife* of the Hebrews. According to the Bible, the Hebrews brought back from Babylon twenty-nine *sacrificial knives* (or perhaps knives for circumcision?) which had been returned by Cyrus.

Weapons have sometimes also had a singular aspect to them. A knife which has disappeared except for a trace in a medieval manuscript is the *becquoysel* (*bird's beak*). Apparently it was a weapon with a hooked blade.

The *kernelers*, or *nut openers*, as their name indicates, were used to remove the kernels of nuts. They barely belong to the knife family since only the end, small in size, resembles a blade while the handle was often more like a key.

Not far in concept from the *secret knives* mentioned in the last chapter, were the *friend's knives* which were in fashion during the reign of Louis XV. These were two knives buttoned together and fitting in a sheath. The knives could be separated and lent to those who were missing theirs.

There are some extreme cases of knives being made that did not even resemble knives. Louis-Philippe *penknives* had the most original shapes — some looked like birds, others like musical instruments. One even had a perpetual calendar. Some had a seal at one end. Knives were sometimes worn as jewelry, especially as charms.

If knives have been disguised as other things, they have also been combined with dangerous weapons. Thus there are, or have been, single or double shot *pistol-knives* and even *revolver-knives* of various calibres. The blade is that of a *folding knife* and opens parallel to the gun barrel, like a bayonet.

Some knives are masterpieces of miniaturization: there are *folding knives* which are much less than an inch long, very easy to hide... and very

These two models were made by Etablissements Thérias and l'Econome in Thiers around 1860. They are farmer's knives, the one at top being particularly for shepherds: it has a lance for bleeding animals. The second is called a 'laguiole' (see glossary); it has a horn handle. Musée des arts et traditions populaires, Paris

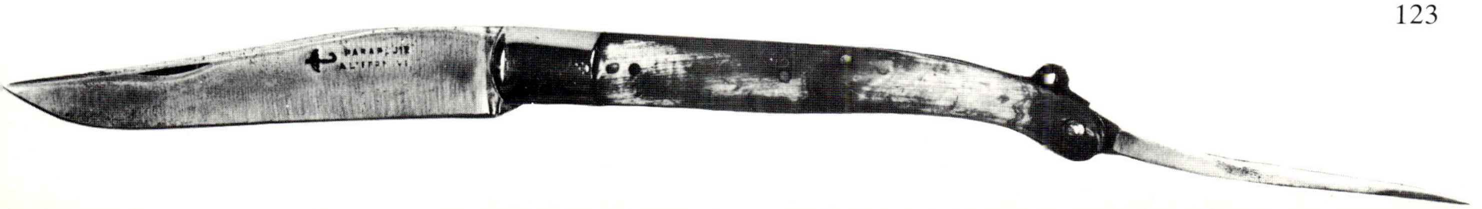

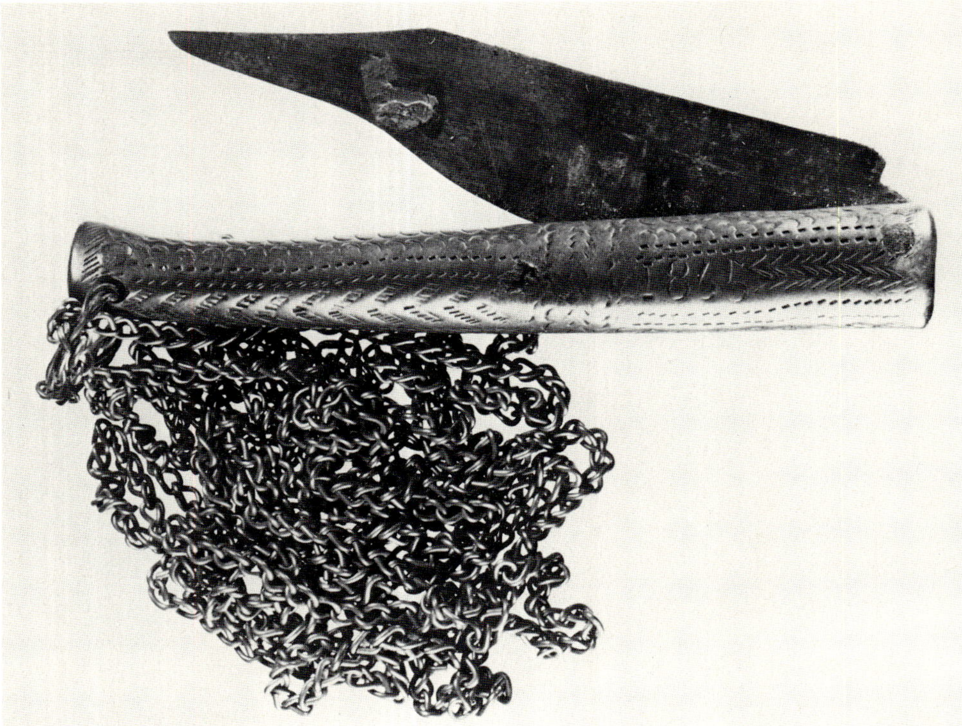

Above left: The handle and the chain of this shepherd's knife, dated 1865, are copper. Massat (Ariège). Musée des arts et traditions populaires, Paris

Above: The handle of this rustic pocket knife is wood. It bears the inscription 'M. Potier'. The blade curves inward both on the inner and outer side. Sarthe, Musée des arts et traditions populaires, Paris

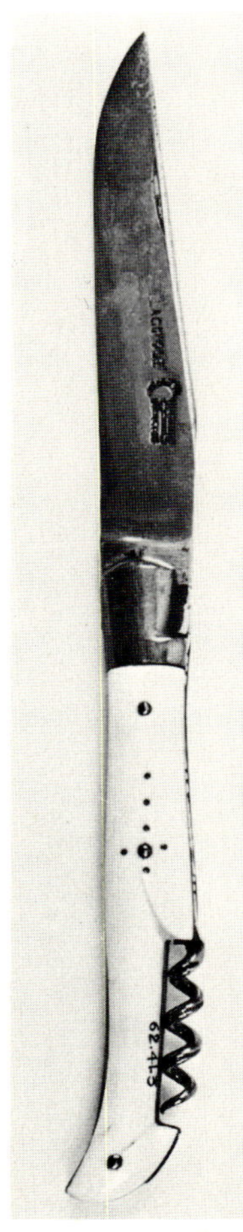

All of these pocket knives appear to be shepherd's models. The first three are springless, the fourth has a punch and a corkscrew. The handle of the first is bullhorn. Millau, France about 1880. Those of the third and fourth are wood, one painted and the other carved. Brittany. The handle of the last one, the most recent, is white bone. Thiers. Musée des arts et traditions populaires, Paris

The fisherman's knife on top is provided with a scaler, which is also a disgorger, and a tiny whetstone for sharpening the fish hook. Length 5 inches. German-made. Below is a knife for yachtsmen. One of the accessories is for tightening knots and the marlinspike is for splicing rope and tightening shackles. All steel. Length closed: 4 inches. English-made. Coutellerie du Mont-Blanc Collection, Geneva

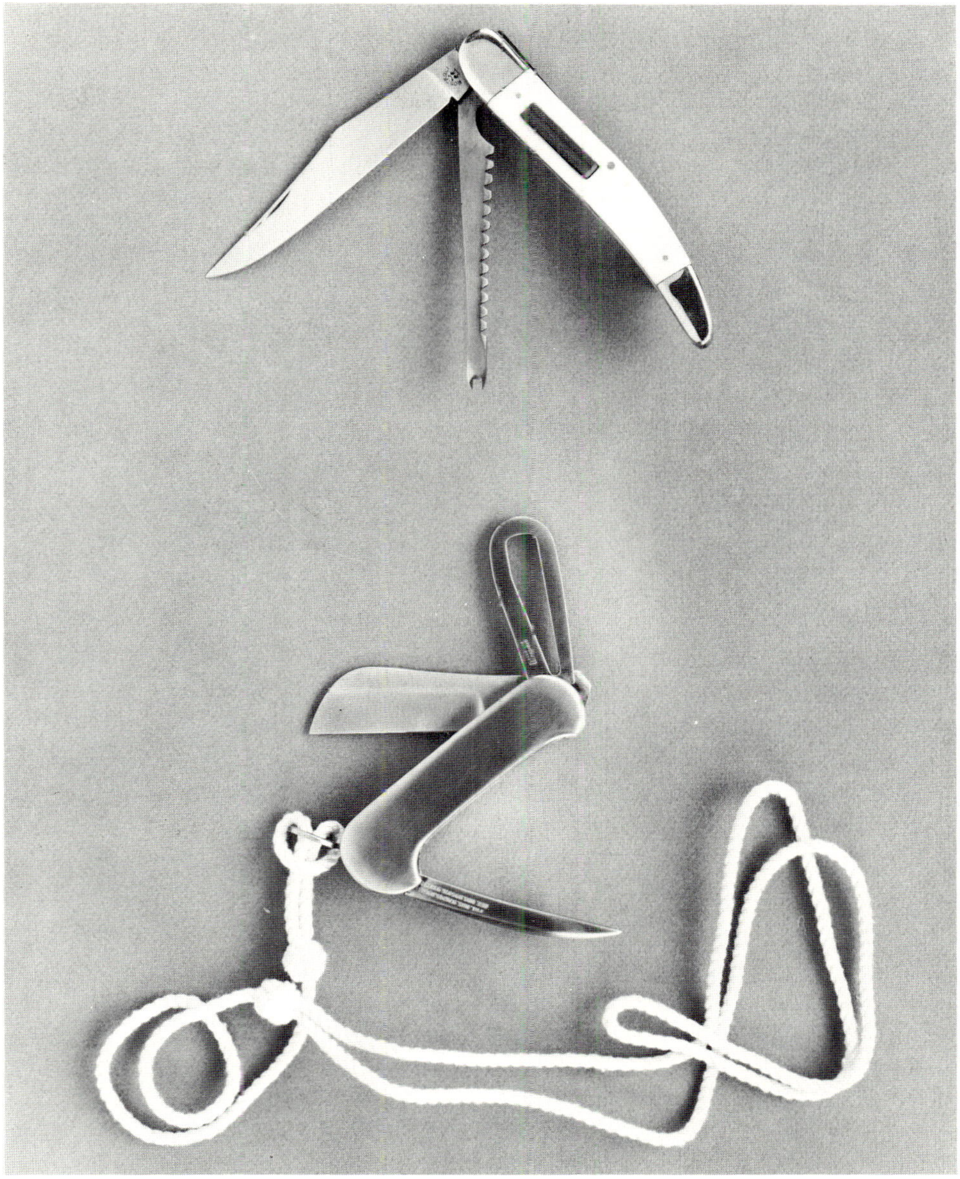

This blade by Case is an astronaut's knife. The alloy is ultra-light. The handle, which is also light-weight, is made of a special plastic and can burn without producing toxic smoke. The end is pierced for a thong. Each model is numbered. Length 16 inches. Coutellerie du Mont-Blanc Coll., Geneva

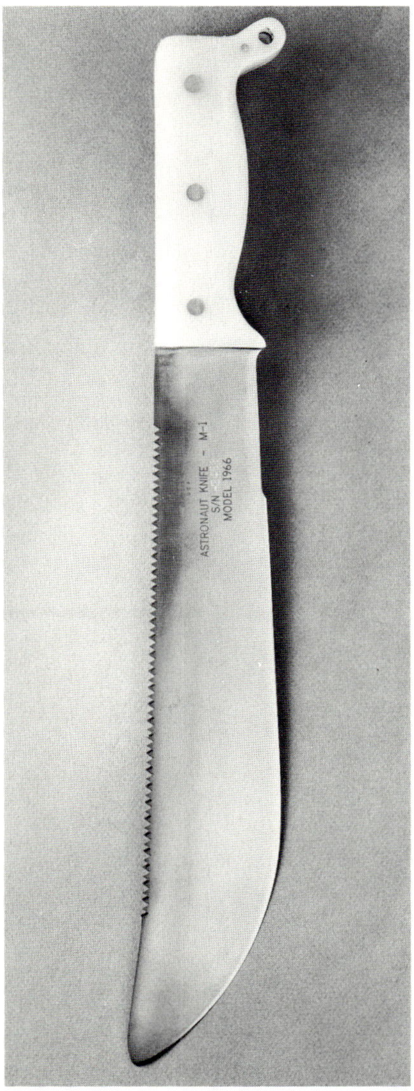

easy to lose. Camille Pagé even mentions some *dwarf knives* of the 19th century of which a hundred could fit into a cherry pit, which seems a bit hard to believe.

Handles can also be unusual. One of the most ancient handle decorations is a stag's hoof. Many hunters had carving sets made using the two front hooves of a deer. The hooves were mounted with shoes and the center of the hoof was inlaid with a shell carved in the form of an animal.

In the 18th century, *hollow knives* could still be found. As their name indicates, the handles were hollow and were used as scabbards for smaller knives, which were sometimes folding models.

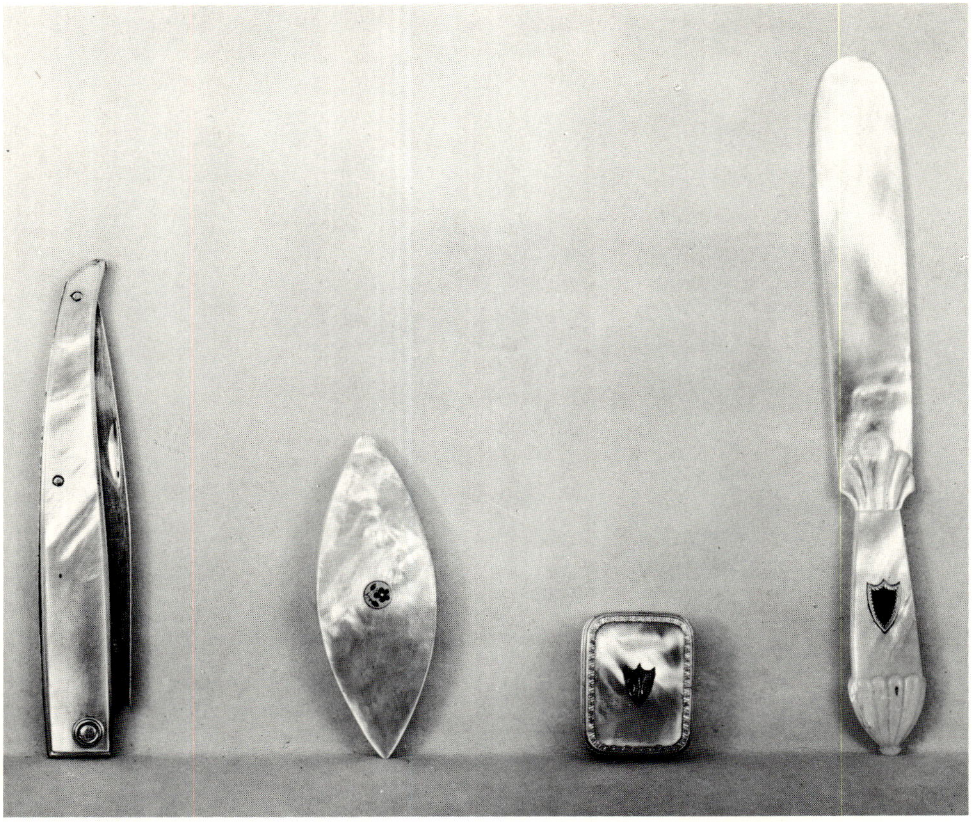

Left: This folding knife and letter-folder belonged to the Empress Marie-Louise. The handle is mother-of-pearl. The work is of Martin-Guillaume Biennais. Schatzkammer der Residenz, Munich

Right: This finely decorated smoker's folding knife has a pair of accessories for cleaning pipe tools. The handle is horn. Length: 9-1/4 inches. French, beginning of the 18th century. City Museum, Sheffield

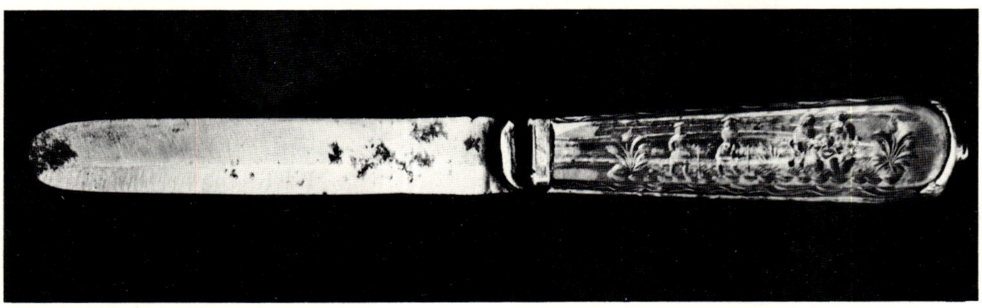

Left: Among ritual knives can be found those used for circumcision (here a Hebrew ritual knife). Various explanations have been given for this rite: hygienic reasons, initiation; but in Judaism it is the outward sign of the covenant between God and Abraham. Circumcision was originally performed with a stone knife. Musée de Cluny

The form of this springless folding knife is unusual: handle and blade together form a curve each in the opposite direction. The handle is ram's horn, the rivets are brass. Length opened: 10 inches. Late 18th or early 19th century. City Museum, Sheffield

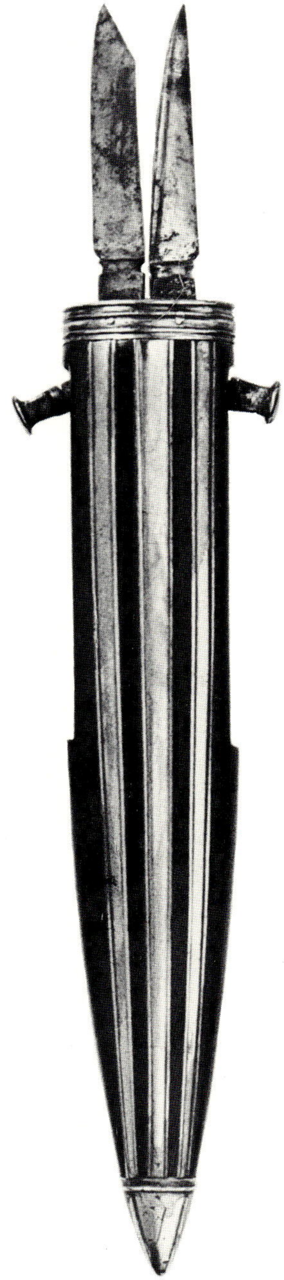

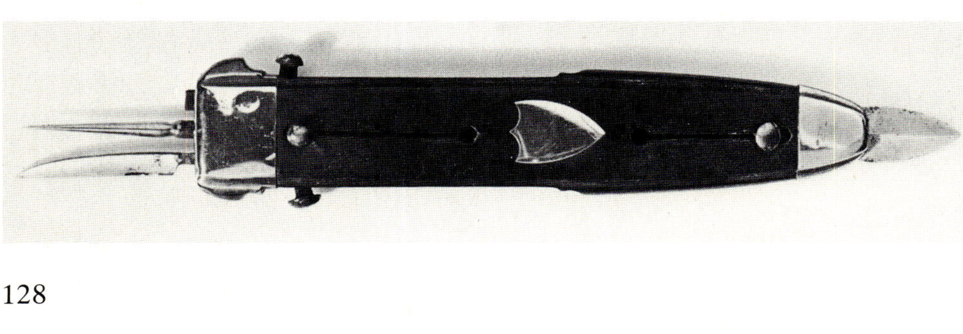

In the beginning, penknives were used primarily for cutting goose quill pens. They were part of the set of writing instruments and were often made in deluxe presentations, even as works of art. All of these penknives are French and date from the 19th century; the two pieces with multiple blades could even go back to the end of the 18th. From left to right and below, the materials used are: mother-of-pearl, enamel, agate, horn and silver (the last two). Total length: between 4-3/4 and 6 inches. City Museum, Sheffield

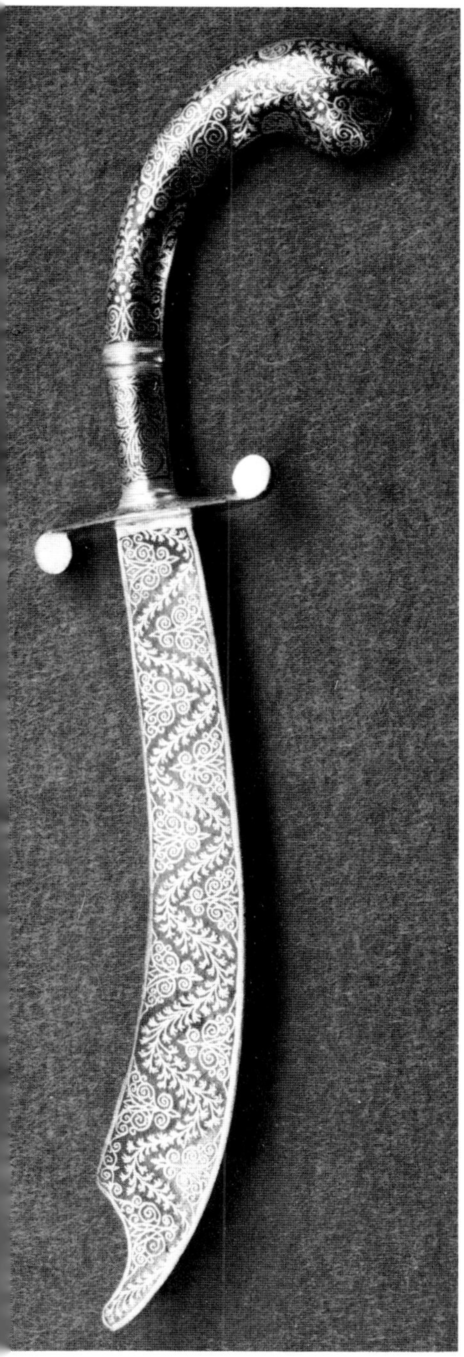

A typical writing set of the 19th century looked like this: an eraser, a seal, an enameled pencil lead holder, a quill and quill sharpener, a scraper (the model shown here is from early 20th century) and a paper cutter. The piece above is Indian 'damascened' work. City Museum, Sheffield

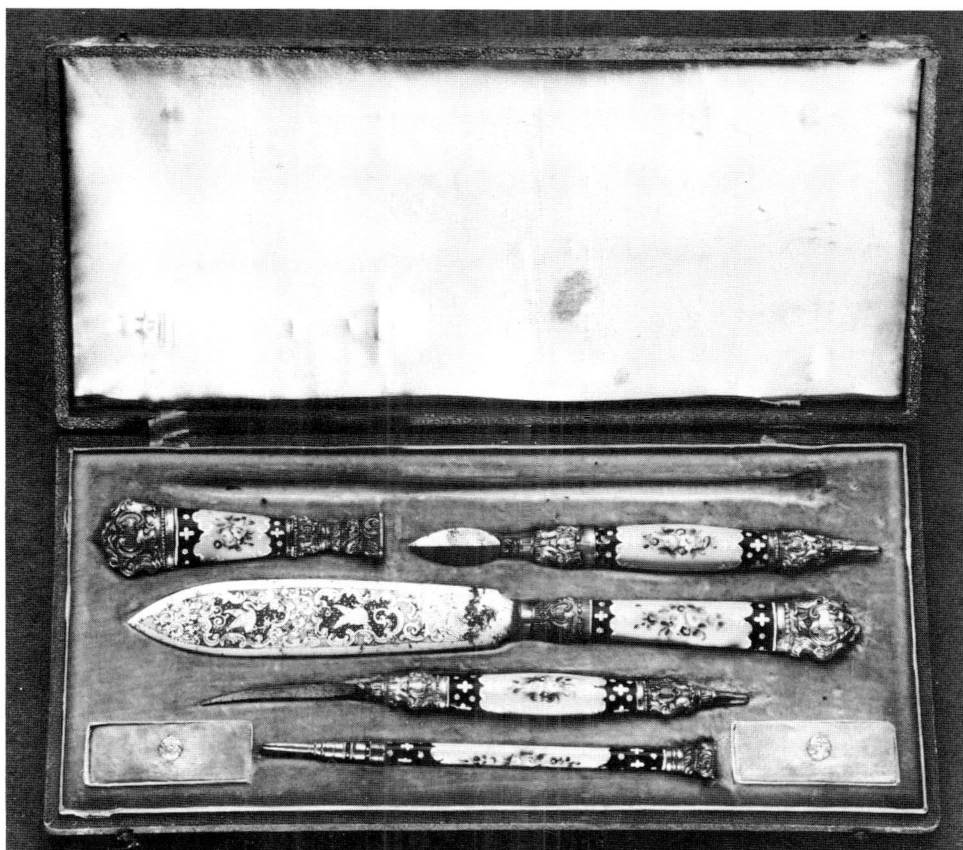

Penknives can also contain a large number of accessories. These three pieces open to reveal a compartment for blades and other accessories, while the base can serve as an inkpot. The one below even has a perpetual calendar and a seal. Length (without the removable blade): 3-1/4 inches. England, beginning of the 19th century. The first penknife is brass, 12 inches, England, 18th-19th century. The second is wood and ivory. 6 inches. France, 18th-19th century. City Museum, Sheffield

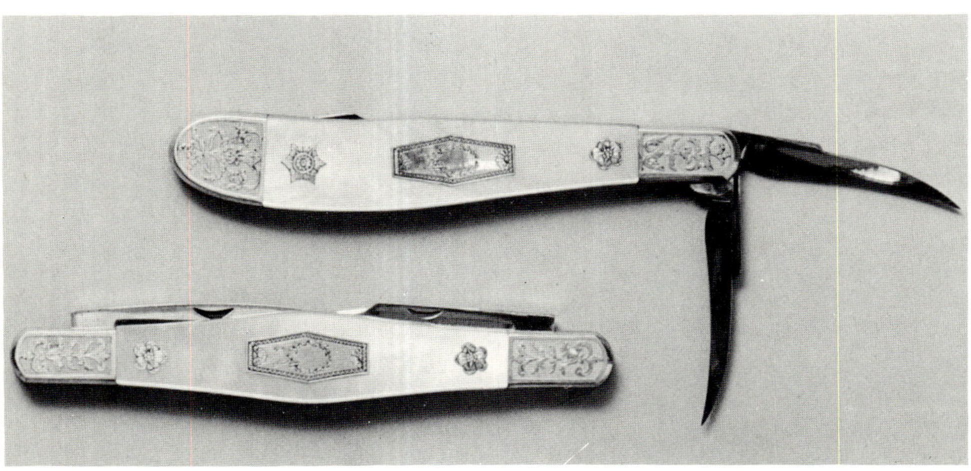

These two penknives, one with two and the other with four blades, are veritable pieces of jewelry. The handle is of mother-of-pearl; the trimmings are silver. Residenz, Munich

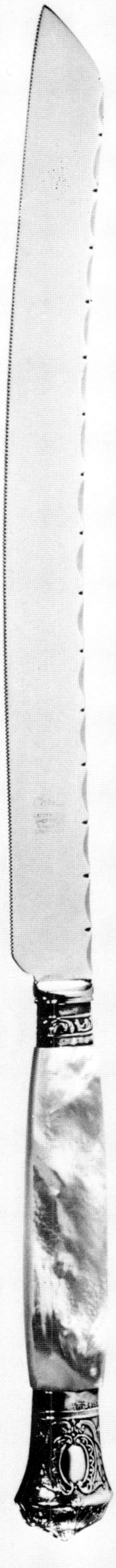

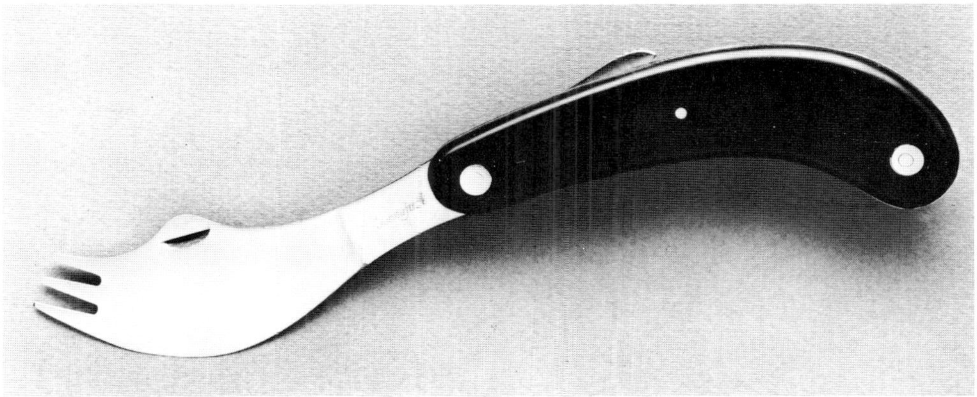

This fork and knife combination with a safety catch has a very unusual shape and was especially designed for the single-handed. By holding it against something it can be opened with one hand. The nail-nick is very prominent for the same reason. Length closed: 4 inches. French-made. Coutellerie du Mont-Blanc Collection, Geneva

Right: This 'coin-knife' by Elci Pernet is silver. Two of the blades fit one on top of the other and the third piece fits on the side. The manufacturer had to adapt the blades to the shape of the handle and the nail-nicks protrude. Coutellerie du Mont-Blanc Collection, Geneva

Left: This 'bride's knife' was used to cut a wedding cake. The handle is in mother-of-pearl and silver. Coutellerie du Mont-Blanc Collection, Geneva

This exhibition piece has a folding cigar-cutter. One of the sides is transparent. Length closed: 3-1/4 inches. German-made. Coutellerie du Mont-Blanc Collection, Geneva

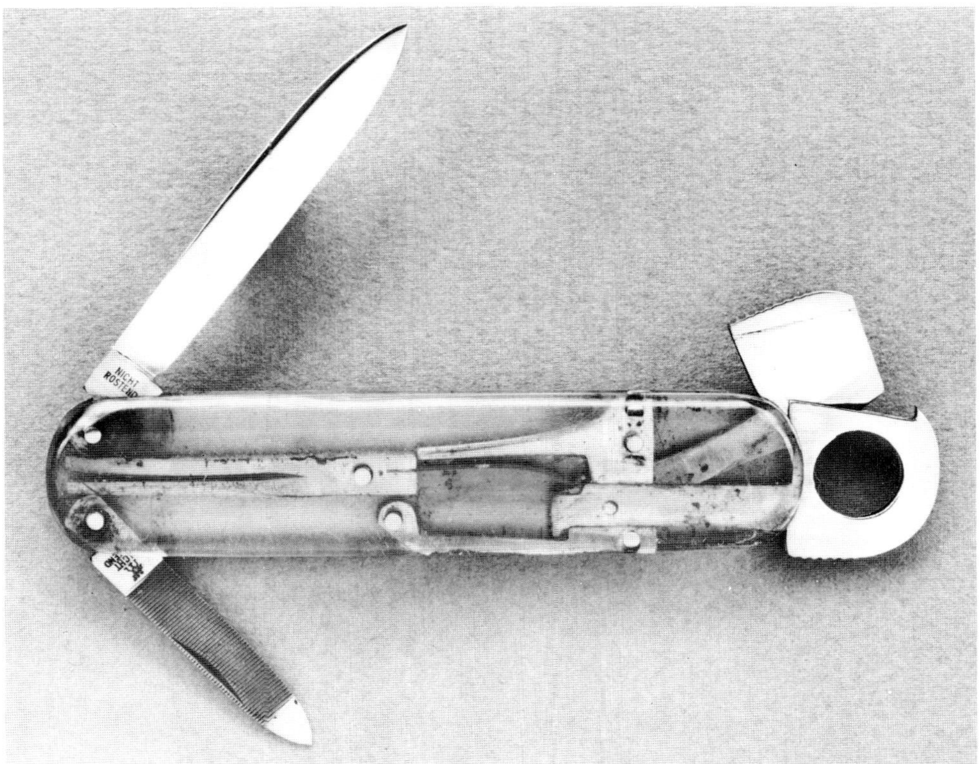

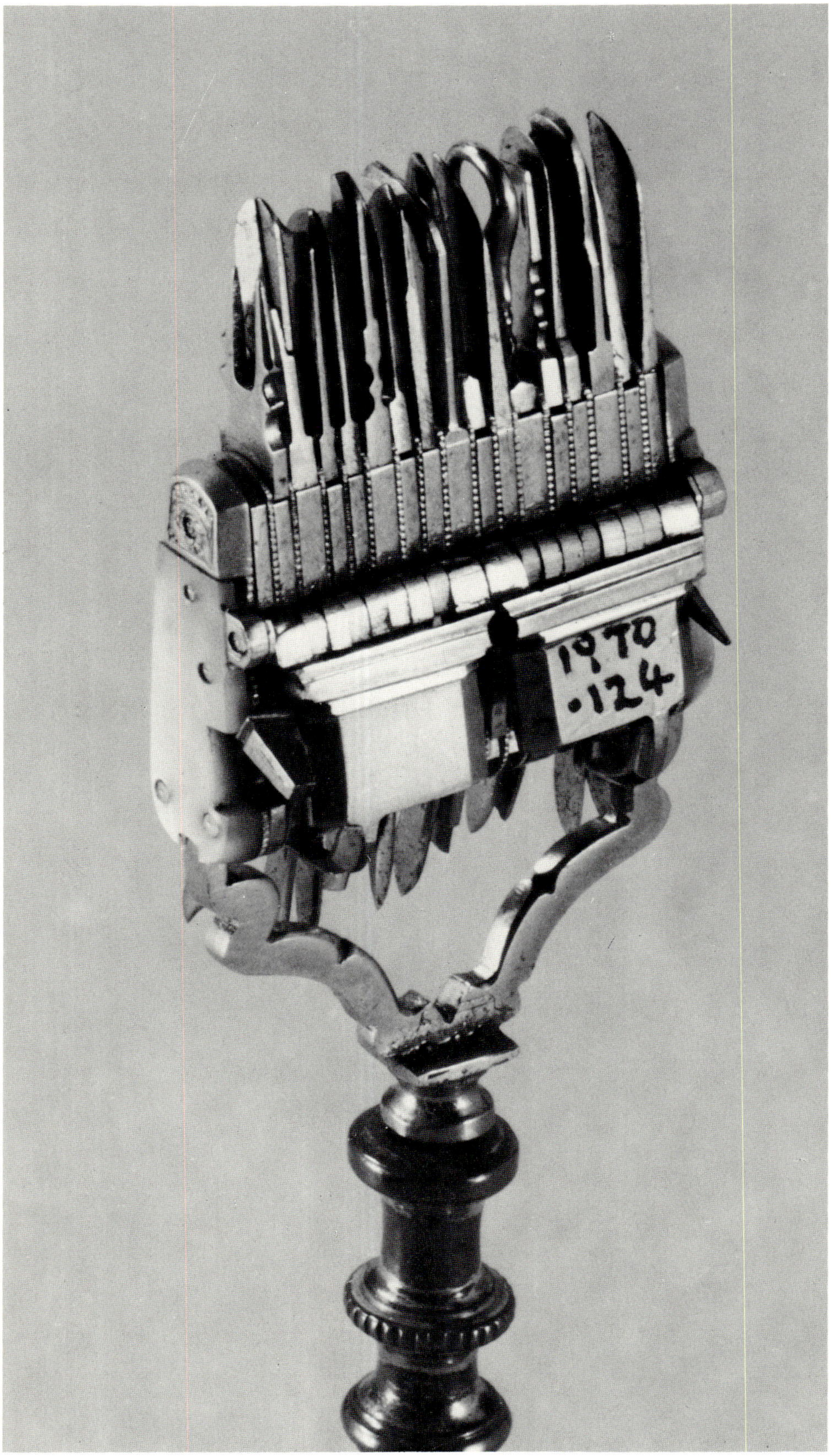

The date and manufacturer of this miniature folding knife is unknown. The knife itself measures no more than an inch long when folded, but it has forty blades and accessories (saws, scissors, fork, etc.). The covers are mother-of-pearl. English, 19th century. City Museum, Sheffield

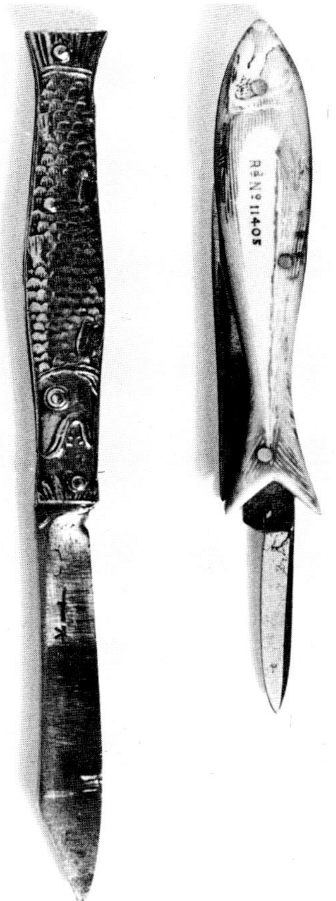

Since they first came into being, folding knives have been made for all kinds of purposes and in all kinds of shapes: fish (the first two), flower and leaves (the third piece from the right) a guitar (next to the last), etc. Below right: Others stand out because of the way they are made, such as the knife with a double handle, or by their function: the third piece from the right is also a snuff box, the one next to it is also a measuring ruler. All the pieces are English, with the exception of the five top right, and date from the 18th-19th centuries. The materials used include mother-of-pearl, brass, tortoise shell and wood. City Museum, Sheffield

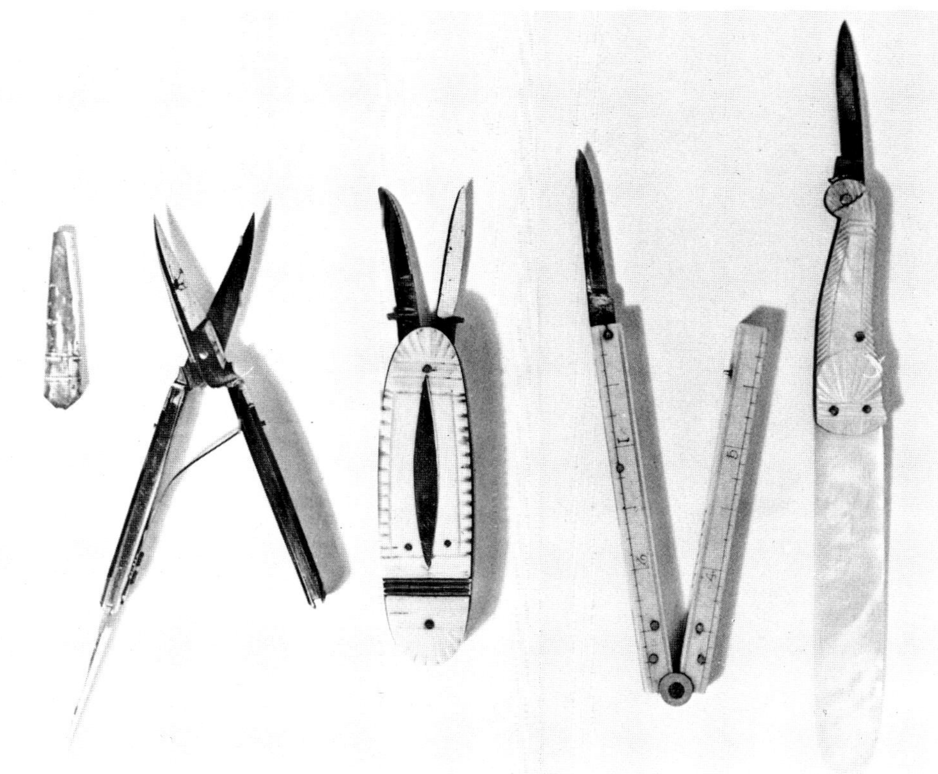

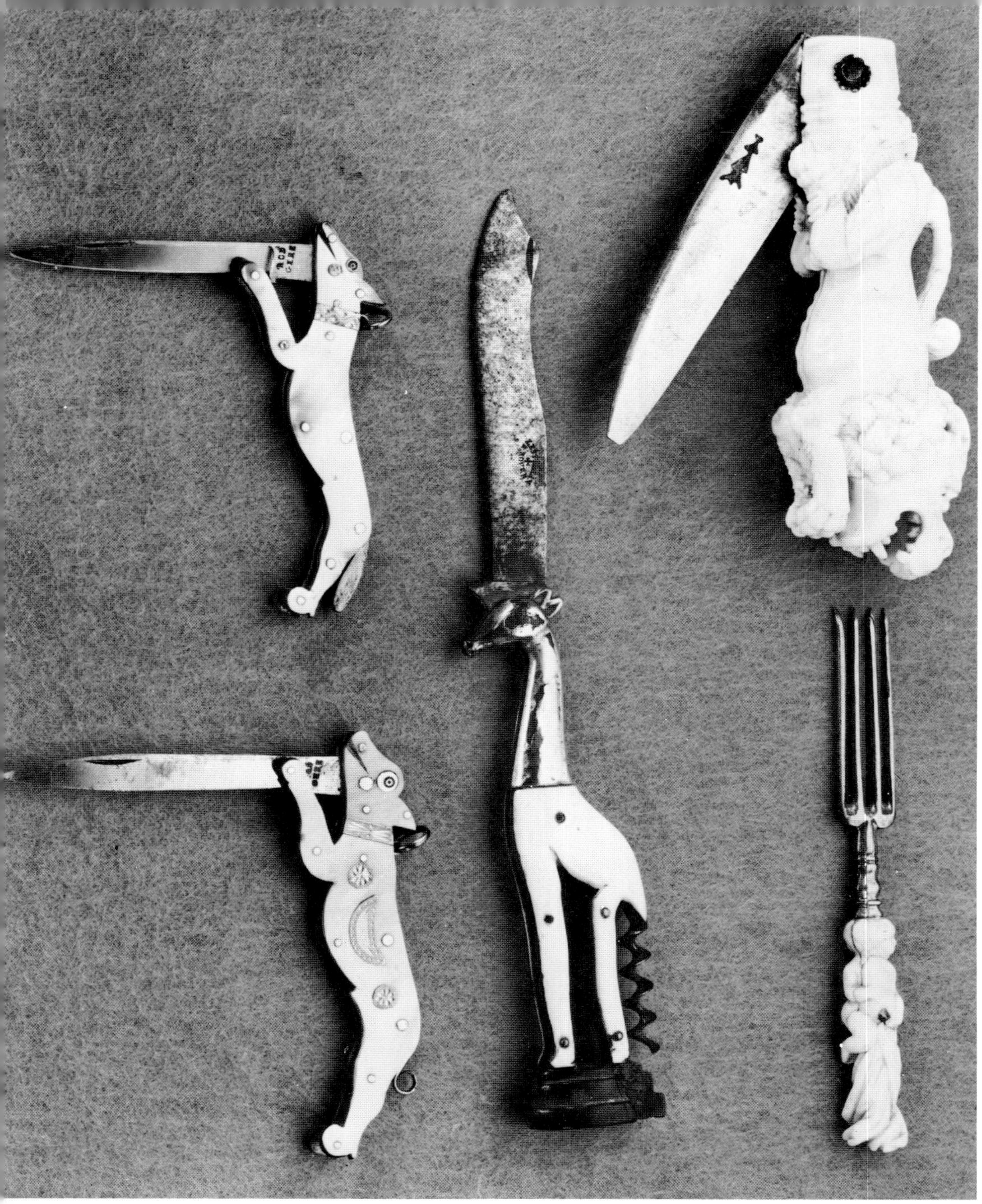

All of these knives with ivory covers are made in the shapes of animals, hunting dogs, a giraffe whose tail is a corkscrew, a lion (the knife) swallowing a child (the fork). English, end of the 18th century. Length closed: 3 inches. France, late 18th century. 4-1/2 inches. Germany, 1680. 6-1/4 inches. City Museum, Sheffield

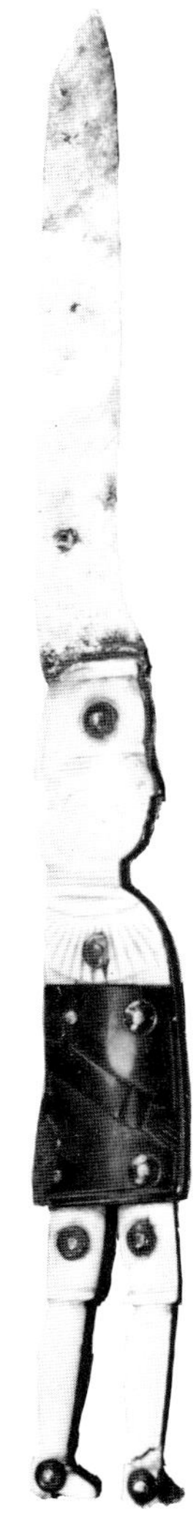

The ivory handle of this knife represents a soldier. Length: 4-1/8 inches. French-made, beginning of the 18th century. City Museum, Sheffield

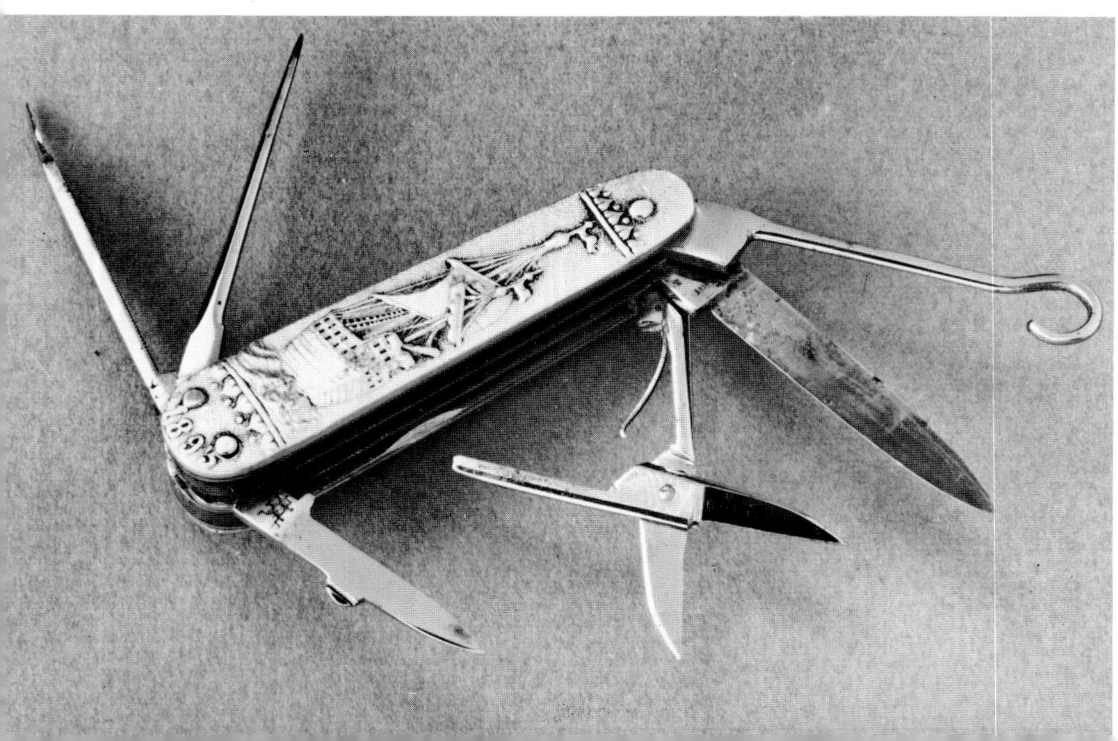

The silver handle of this knife of the Henckels brand has the caravelle 'Santa Maria' of Christopher Columbus on one side and a portrait of the discoverer on the other. It has two dates: 1892 (400th anniversary of the discovery of America) and 1895. Length closed 2 inches. German manufacture. Coutellerie du Mont-Blanc Coll., Geneva

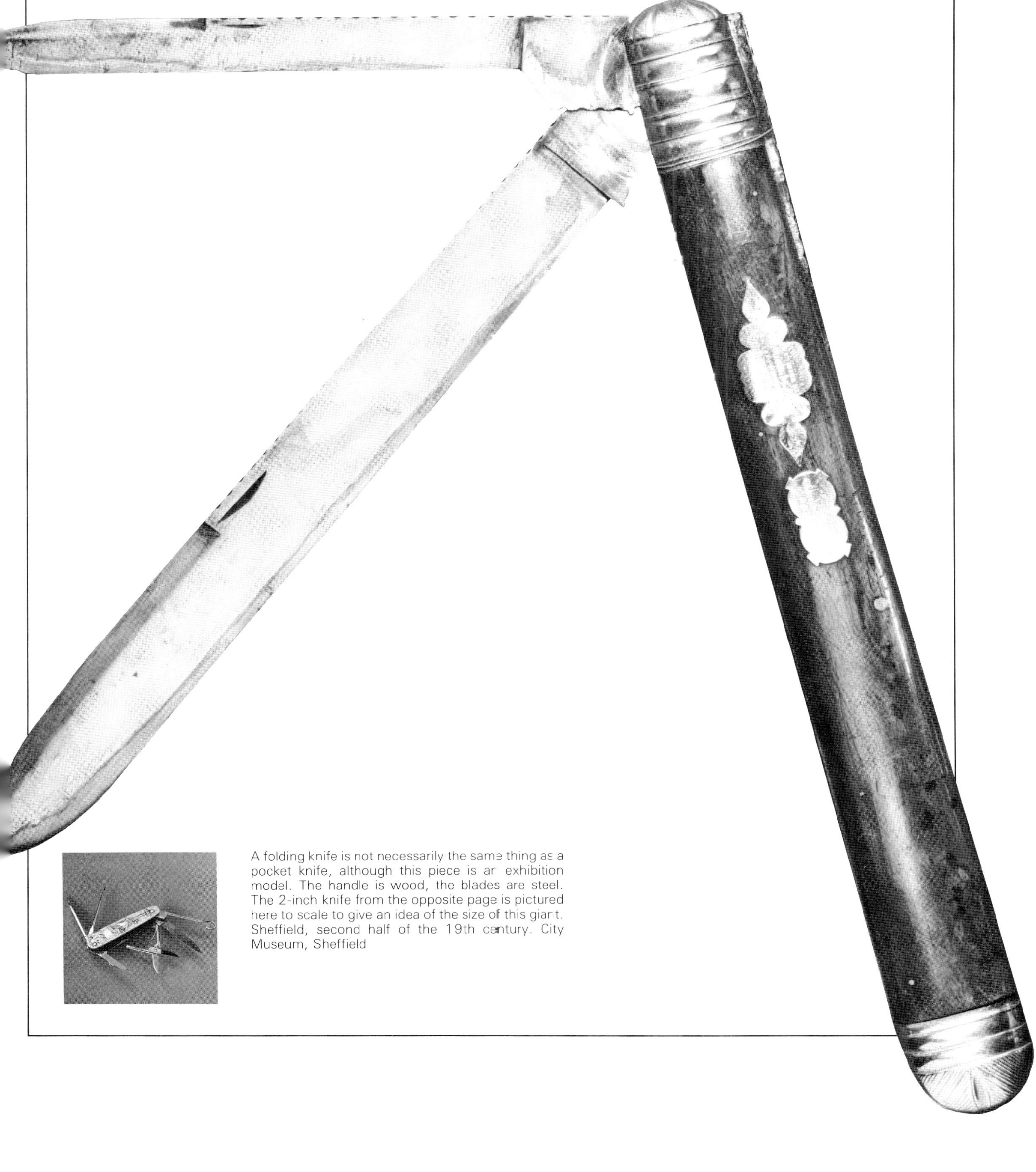

A folding knife is not necessarily the same thing as a pocket knife, although this piece is an exhibition model. The handle is wood, the blades are steel. The 2-inch knife from the opposite page is pictured here to scale to give an idea of the size of this giant. Sheffield, second half of the 19th century. City Museum, Sheffield

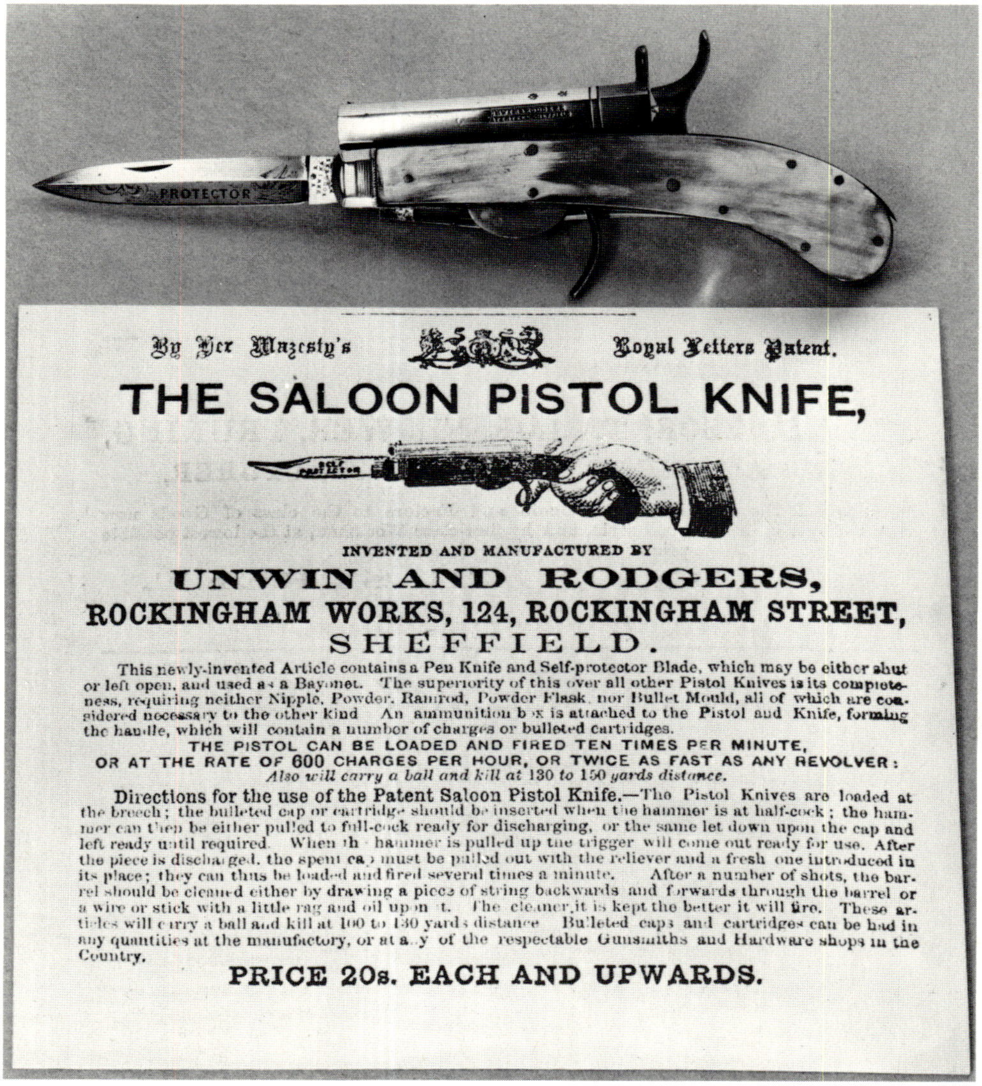

By Her Majesty's Royal Letters Patent.

THE SALOON PISTOL KNIFE,

INVENTED AND MANUFACTURED BY

UNWIN AND RODGERS,

ROCKINGHAM WORKS, 124, ROCKINGHAM STREET,

SHEFFIELD.

This newly-invented Article contains a Pen Knife and Self-protector Blade, which may be either shut or left open, and used as a Bayonet. The superiority of this over all other Pistol Knives is its completeness, requiring neither Nipple, Powder, Ramrod, Powder Flask, nor Bullet Mould, all of which are considered necessary to the other kind An ammunition box is attached to the Pistol and Knife, forming the handle, which will contain a number of charges or bulleted cartridges.

THE PISTOL CAN BE LOADED AND FIRED TEN TIMES PER MINUTE,
OR AT THE RATE OF 600 CHARGES PER HOUR, OR TWICE AS FAST AS ANY REVOLVER:
Also will carry a ball and kill at 130 to 150 yards distance.

Directions for the use of the Patent Saloon Pistol Knife.—The Pistol Knives are loaded at the breech; the bulleted cap or cartridge should be inserted when the hammer is at half-cock; the hammer can then be either pulled to full-cock ready for discharging, or the same let down upon the cap and left ready until required. When the hammer is pulled up the trigger will come out ready for use. After the piece is discharged, the spent cap must be pulled out with the reliever and a fresh one introduced in its place; they can thus be loaded and fired several times a minute. After a number of shots, the barrel should be cleaned either by drawing a piece of string backwards and forwards through the barrel or a wire or stick with a little rag and oil upon it. The cleaner it is kept the better it will fire. These articles will carry a ball and kill at 100 to 130 yards distance. Bulleted caps and cartridges can be had in any quantities at the manufactory, or at any of the respectable Gunsmiths and Hardware shops in the Country.

PRICE 20s. EACH AND UPWARDS.

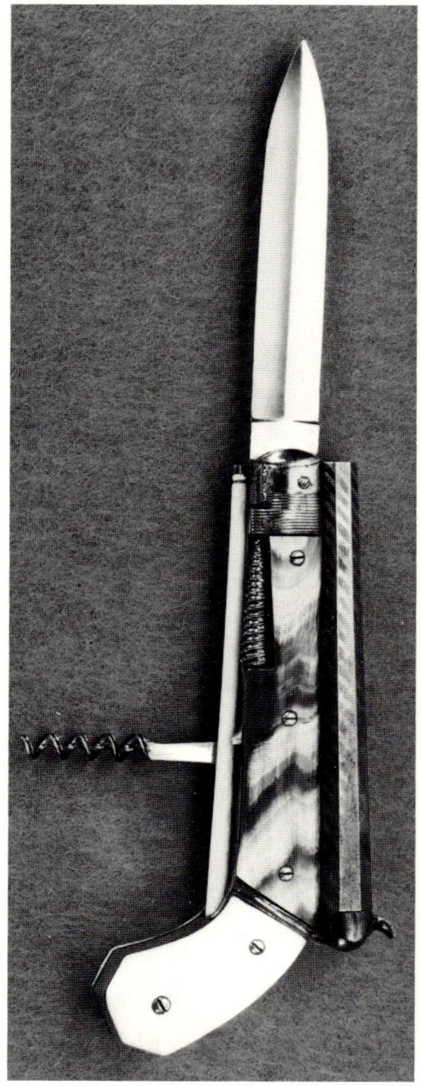

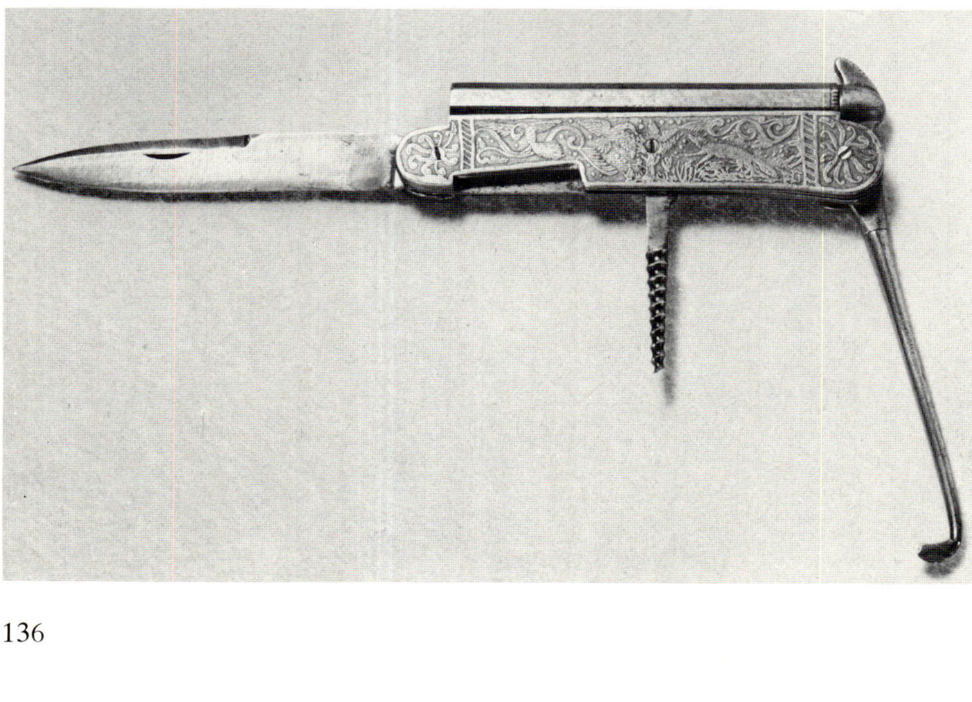

'Pistol knives' were popular in the last century, and although easily concealed were not very accurate. The triggers fold away. In the double-barreled weapon above, there are two triggers which emerge when the pistol is cocked. The knife springs out when one is pulled. Length: 10-1/2 inches; caliber 6.35 mm; Sheffield, around 1870. The stock of the second is ivory. 9-3/4 inches; caliber 6 mm; France, around 1850. In the last model, a horse's hoof pick is the butt. Length: 9-1/2 inches, caliber 9 mm. France or Belgium, 19th century. City Museum, Sheffield

Right: The 'Sheffield Knife' is described in the 'Guiness Book of Records' as the knife with the greatest number of blades in the world, one for each year of the Christian Era.It will be considered complete in the year 2000. Many blades are marked with a date. Height: 31-3/4 inches. Original manufacturer and owner: Joseph Rodgers & Sons Ltd. (since 1822). Since 1969: Stanley Tools Ltd., Sheffield

era. Acquired by Stanley Works (G.B.) Ltd., Sheffield, in 1969.

Clockwise from top: the first four are French four-piece manicure knives. The handle of the first (tortoise shell) and the second (ivory) are approximately triangular. The other two (shell and ivory, respectively) are oblong. The next miniature, a five-piece knife, has hand-engraved scales. Both scissors fold flat as does the file in the first knife. This model is of American manufacture. The last knife is a four-piece French manicure knife with a tortoise shell handle. Coutellerie du Mont-Blanc Collection, Geneva

These six knives, with bolsters and wooden handles, fit into a gold-reinforced nutshell. They can open and the blades work. (The museum of Sheffield has a nutshell with twelve knives). The match gives the scale. French-made. Coutellerie du Mont-Blanc Collection, Geneva

The dimensions of this case (4 × 2 inches) give an idea of the smallness of this display table service, a deluxe set. The handles are ivory. Nothing is missing, not even the carving set. City Museum, Sheffield

Who would guess that this belt contains a 'Bowen push dagger'? The blade slides into a sheath on the inside of the belt. Its proprietors consider this a survival knife which is always at hand, without risk of wounding the wearer. The blade is decorated with a landscape by Dillingham. American-made. Coutellerie du Mont-Blanc Collection, Geneva

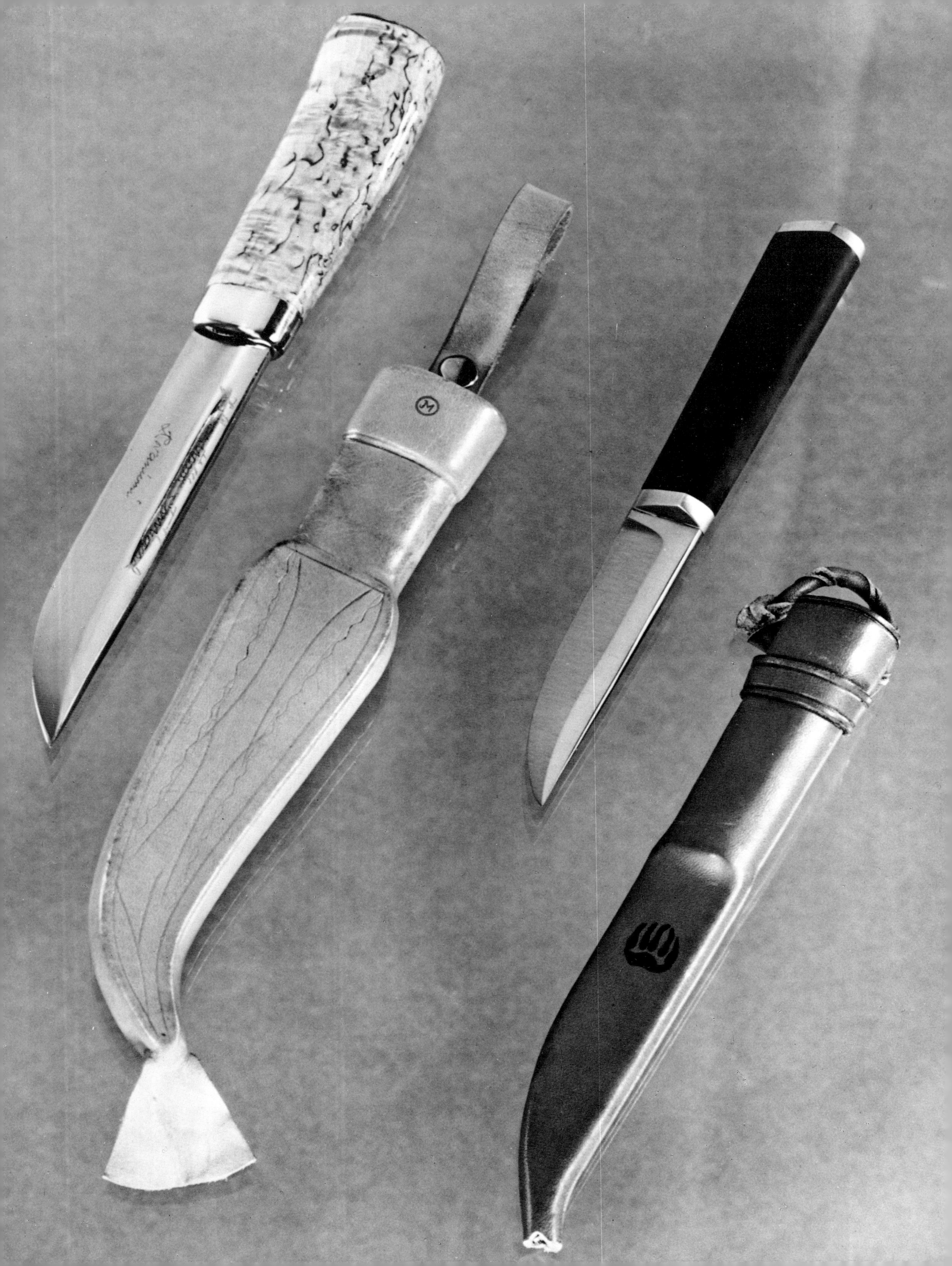

The exotic can often be found right on our own doorsteps. And sometimes a knife with a particularly evocative name is simply an ordinary word for knife in a foreign language, designating nothing more than the everyday concept familiar to us.

The Islamo-Arabic world has long been noted for the quality and beauty of its blades. Sweeping generalizations should not be made, however, if only because the art of Islam presents such variety and is dispersed in time as well as in space. Moreover, knowledge of the metals used is very meagre. Few very old pieces have survived, but the superiority of Near Eastern arms was recognized up into the Ottoman age. While some manufacturing processes antedated the age of Islam, they were nevertheless brought to perfection during it. This applies to 'Damascus' blades, which were of Indian origin, and which can be traced back at least until the age of Diocletian. This also applies to the Toledo steels, whose origin may be older than the Roman period. As for the curved blade, it is clearly of Turko-Mongolian origin.

It has often been said that in the Islamic world a hand weapon is masculine jewelry. This is certainly the case with finely ornamented daggers, decorated with arabesques, inscriptions, and sometimes with a magic square. But in addition, a hand weapon has a symbolic value, and the images of animals which were often used to decorate side arms must be seen in this context. Moreover, the hunt was considered as a school for war, even though it was often practiced in reserves.

The arts of India have been particularly marked by Islam, above all in the North and beginning with the Mogul period (16th century). But as has been seen in the first chapter, the Indians possessed a remarkable tradition of steel forging going far back in time. A typical Indian knife is the *kuttar* (or *kattar*), with a triangular blade and a thick point, some two inches wide at the base and a foot long. This *tiger knife* was mounted on an iron bar which bends back at right angles at both ends. Parallel to the bar on which the blade is mounted, is another bar which joins the two bent lengths. The weapon is grasped by this second bar and manipulated by the whole arm. The steel of the *kuttars* is often finely worked, and the overall effect of their decoration is impressive. In some *kuttars* the grip is covered by a protective hand cup. This cup can itself be decorated, sometimes with vine scrolls. In the Kutch (West India), there are knives meant for cutting betel nuts, which are used as a dye. *Betel knives* also exist in several areas of South Asia, particularly in Indochina. Until recently the steel used was wootz, or India steel, a case-hardened and cast variety obtained from a magnetic oxide that contains up to 72 per cent iron. Once the steel is made it is bleached by a grinding wheel, after which it is plunged into an acid bath which brings out a damascened texture.

Chinese metal working is particularly ancient since its Bronze Age goes back to 1600 B.C. One category of tools, which is very widespread in the Far East, is the *rectangular knife*, which is sometimes also trapezoidal, semicircular or crescent shaped. A variety of it can be found among Paleo-Siberian peoples, Eskimos, and even in pre-Colombian Indian remains in New York State. The conservative nature of the Chinese can be clearly seen from the fact that an iron version of this knife was still used in recent times in North China. Another typical form was the *hatchet-dagger*. This term designates a weapon which has both a point and an edge. Here again the relative stability of a form persisting through at least 2,500 years is remarkable. Examples

Morocco, at the farthest reach of the western Islamic world, did not undergo Ottoman domination and its art displays an affinity for the archaic. The country has been known for centuries for its daggers. The model above was made in the 18th century and antedates the decay of Islamic art. It comes from Tagmut in the Anti-Atlas. The materials used are silver-plated steel, brass and wood. 16-3/4 inches. Musée des arts africains et océaniens, Paris

With variations in shape and size, the 'puukko' is the typical knife of the Finns and Lapps. That on the left (by Marttiini-Rovaniemi) is an example of a more classic design with a blood groove. The handle is birch. The sheath is partly wood. Collection of Jean-Rodolphe Piccard, Lausanne. The knife on the right is a modern interpretation of a traditional model by Tappio Wirkkala. The hilt is brass and the handle plastic, impervious to extreme temperatures. Coutellerie du Mont-Blanc Collection, Geneva

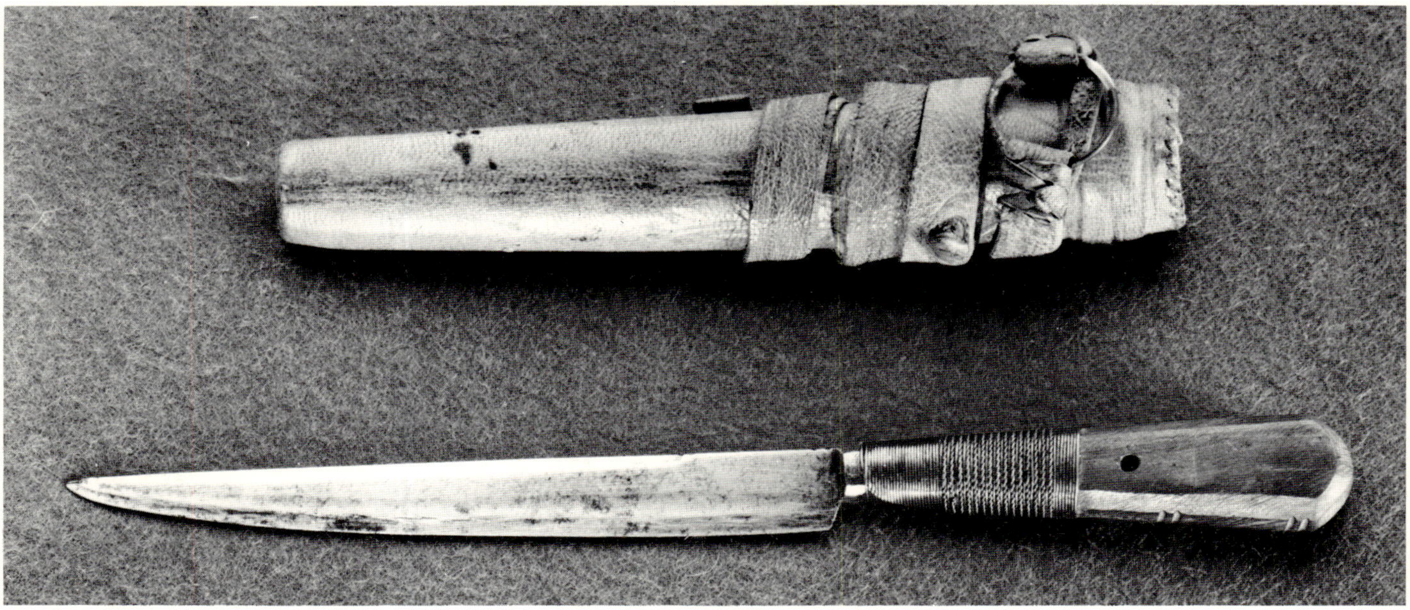

The 'khodmi' with an iron blade is of Moroccan origin. The handle is wood wrapped in brass wire; the sheath is leather. Length of the knife: 10 inches. 19th century, City Museum, Sheffield

made of both bronze and jade have been found. More recent knives, which resemble more familiar models, will not be mentioned here.

In Japan, up until recently, the armorer was held in great esteem. This particular status of the blacksmith can be found in other parts of the world, as far as the Sudano-Guinean region. The Japanese swordsmith prepared himself for his work with prayer, great purity of mind, and absolute chastity. The making of a blade could take months or even years. But all this care was reflected in the resulting quality — in the last century Japanese blades equaled the best European ones. They were remarkably regular and the edge was bevelled. The handle was a simple piece of soft wood into which the blade was thrust. The *fuchi* or bolster band could be interior or exterior. The handle was decorated by plain twisted rushes.

The pruning hook with a wooden handle and serrated edge was used for gathering dates. Total length: 16 inches. Tunisia. Coutellerie du Mont-Blanc Collection, Geneva

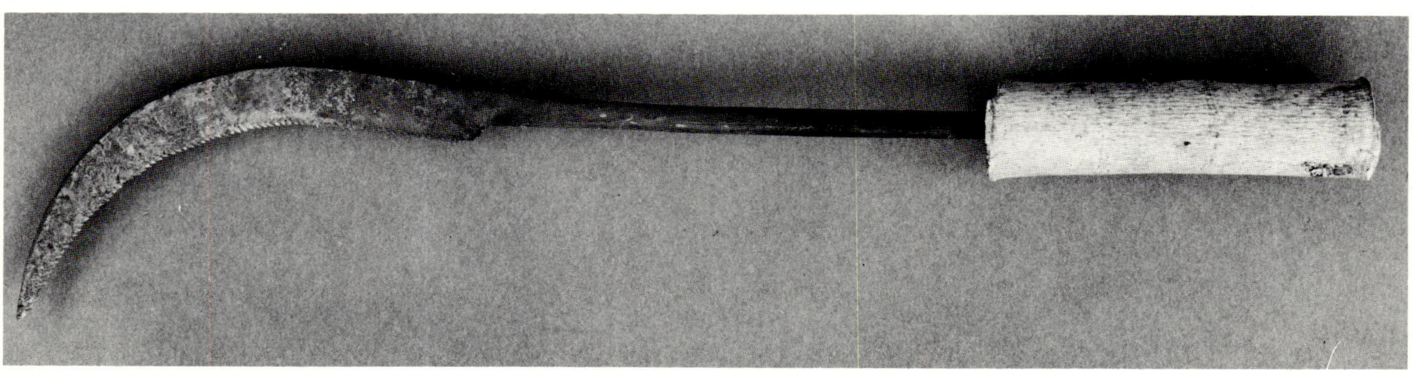

Africa has almost as many styles of knives as it has ethnic groups, as can be seen by comparing these three knives of a single country, Zaire. The first is of Kuba (Kasai) origin, the second comes from the tribe of the Sungu (also Kasai), and the last is from the Ngala, a confederation of peoples around the midpoint of the Zaire River. The blade of the weapon on the left is blackened iron and swells in the direction of the point; on both sides it has a central ridge and grooves. The handle, of light wood, covers the heel of the blade. Note the large ringed pommel. The tang has, like the other three knives, a copper stud at the end of the pommel. It is a dress or fighting weapon of the last century. Length: 13-1/2 inches. The blade of the central knife is worked with light ridges. The spiral handle, of tan wood, is decorated with sheets of nailed copper. The large cap is also copper. 12 inches. The blade of the dagger on the right ends in a curious ogival form, looking like the snout of a pike. The two sides of the blade are decorated with stripes and crosses. The tan wood handle has a conical pommel covered with copper leaf, trimmed at its base with the same material. This 'dress weapon' probably dates from the 19th century. Length: 15-3/4 inches. Musée d'ethnographie, Geneva

Secondary cultures

Removed from the great centers of civilization of which we have been speaking, there are a number of regions with more or less indigenous cultures, which have nonetheless undergone influences from their more powerful neighbors.

Before the present times, Korean knives were only slightly tempered or not at all, so that their cutting effectiveness might well be questioned. Tibetan iron, on the other hand, was held to be of better quality than the Chinese. The Tibetans had knives of different sizes which they wore at the belt. One of them, the *phur-bu*, a 'ghost knife' or 'devil dagger', was used for exorcism ceremonies.

The only common ground of the various peoples of Indochina is that they all underwent the same cultural influence by their two giant neighbors and belonged to the same colonial power. Speaking of Tonkin in the last century in *Le Tour du monde*, Dr. Hocquard expresses his admiration for the makers of handles who work with precious woods inlaid with mother-of-pearl. "The workers... are true artists. The wood used for the inlay is of two kinds: either a rosewood... or a very rare variety of ebony. After being assembled by the wood worker [the object] to be inlaid is passed over to an artist who draws a design for the mother-of-pearl decorations on tracing paper... the craftsman traces the designs... and choses the mother-of-pearl which he requires... those who do the inlay cut the mother-of-pearl — with the most beautiful highlights from emerald green to rose — into small pieces about an inch square. They have a real talent for combining different colours in their inlays." The other operations include polishing, the actual inlaying and glueing.

The principal knife of the Annamese is known as the 'cut-cut', a sort of cutlass or machete used for many things... including piracy. A similar instrument was used in Cambodia, where the extraction of iron was the activity of tribes which were considered 'primitive'.

In Thailand, various knives were used, one of which was meant for cutting the areca nut. Since the juice corrodes iron and even steel, a bronze knife was used for this purpose, eight inches long, with an ebony handle. The Laos, who are related to the Thais, imported their blades from China but made the handles themselves out of artfully chased silver, and the sheaths out of wood. As often happens with knives of the Far East, the final result resembled a club more than a cutting instrument.

The national weapon of Burma is the *dah*, which is a universal tool for all household tasks, including cutting trees, working wood and clearing the bush. Its length varies between sixteen and thirty-two inches; it does not have a guard, and the wooden handle is attached by two copper bolster

The Humbi are a Bantu people from southwest Angola. This dagger made by them has a wooden handle covered with frets of red and yellow copper. The blade, on the right, is iron and has a full length tang. The wooden scabbard is also covered with alternate red and yellow copper frets, and has a large chape in the form of a half-moon. The leather tong attached the weapon to the belt. 14-1/4 inches. Musée d'ethnographie, Geneva

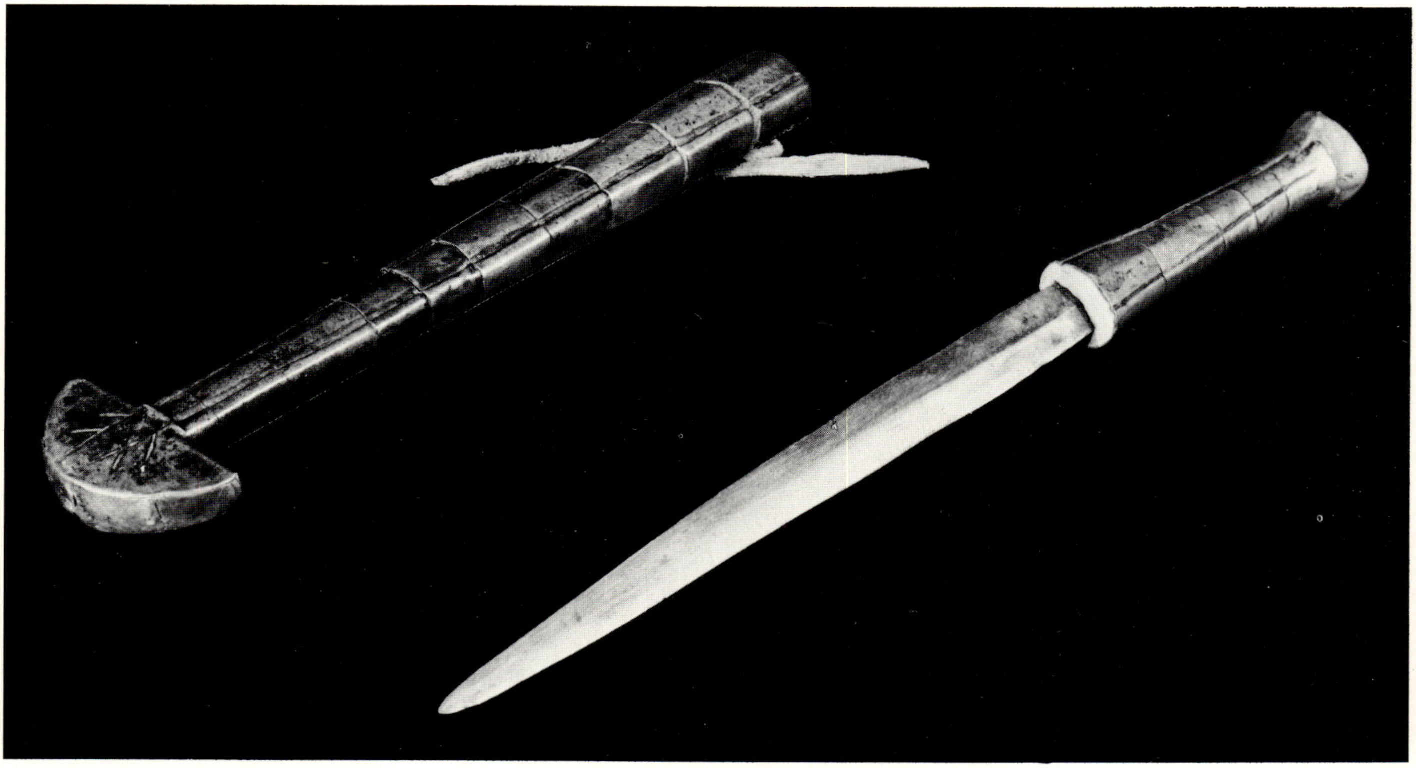

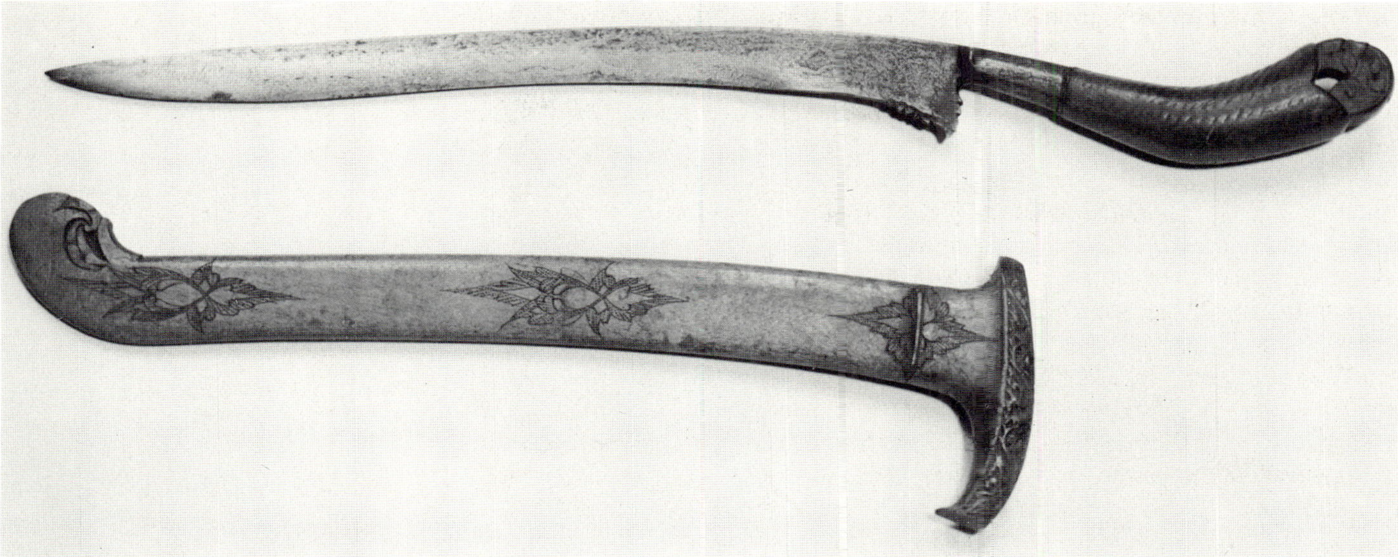

This weapon resembles a 'yatagan' because of the S-curve. However, it is an Indonesian dagger (Atjeh, Sumatra). The handle is horn, the scabbard is wood decorated with carved floral motifs. Length: 16-1/4 inches. Musée d'ethnographie, Geneva

bands. The end of the blade is square, but longer models with a pointed end also exist.

The most famous Nepalese knife is the *kukri* which is generally accepted as being of good quality. It is a kind of cutlass with a curved blade that gets wider towards the point. It is kept in a leather sheath, which also holds two small knives.

In the Malay world (Malaysia, Indonesia and the Philippines) there is the greatest variety of knives in all Asia. It is here too, that the ethnic diversity is also greatest. The favorite weapon of the Malays is the *kriss*, a formidable sixteen-inch dagger which in earlier days never left its owner's side. Usually, it was worn in the belt with the handle sticking out to the right. The wavy blade begins wide and narrows towards the point. At the top, on the side, the *kriss* has a sharp appendage whose purpose is unknown to us. The tang, which is sometimes perpendicular to the blade, goes through the length of the handle. The handle does not have a guard. After being tempered, the blade is kept for ten days or so in sea water. The oxidation is further encouraged by the use of acidic lemon juices giving the surface a moiré finish.

In Malacca, a knife called a *lading* can still be seen. It is used to cut grass and underbrush. The *golok* — heavier at the point than at the handle — is used for removing the bark from trees; the *pisow-raut* is a knife for butchering; the *kachik* for cutting betel nuts. Many of these knives are to be found in the Malay archipelago, such as the *pisow-raut* of Java whose blade resembles that of a Western *butcher's knife*. This knife does not have a guard, but has a scabbard of impressive size. Without any other tool than the *bendo* (a knife resembling the *golok*) and the *pisow-raut*, the Javanese built their homes out of bamboo poles, palm leaves, and coconut fibers. The dayaks of Borneo have several kinds of daggers whos handles are decorated with animal hair. The Tagals of the Philippines use several different kinds of cutlasses which vary in length and thickness of the blade. The handle is made with an iron chape at each end. The Malays have a kind of sword dance called the *kalifa* which consists of turns executed to the flourishes of daggers and other arms.

The knives of so-called 'primitive' peoples

A very ethnocentric mentality has often made Europeans think that the peoples of other continents who are out of the mainstream are 'primitives' who must be 'civilized' at any cost, even that of their own culture. And yet, even if consideration is limited to knives, it can be seen that much of what these peoples have made in the 20th century can stand comparison with Western knives, both for beauty and for originality. From the criterion of function, most of them answer perfectly the needs of those who use them.

Although the peoples of central Asia and southern Siberia have been exposed to outside Moslem — and occasionally to Russian — influences, this cannot be said of the inhabitants further north, who have in many ways remained in a more uninfluenced state. Until recently the Samoyeds buried the knives and other objects of their dead along with them, which reflects their belief in a life after death. The Yakuts are the Turkic people who are settled furthest north, and they have kept the essentials of their shamanist religion alive as well as the technique of making knives out of walrus teeth, although as early as the 18th century they learned the use of iron.

The Chukchi, who before the October Revolution were satisfied with their slings and rough bows, like the Yakuts used knives out of bone and walrus ivory, which had a marble-white appearance. Most paleo-Siberian groups shared similar cultural circumstances as those described above. But, as has been seen, some of them were acquainted with knives of Chinese origin, or had evolved similar ones of their own. Most of the peoples of central Siberia have known the use of iron, extracted in ways similar to those used by early Western people for at least several centuries. This is true of the Ostyaks and the Kalmyks and other Mongol people as well as the Tungouzes. Many of them carry a whetstone to keep their blades sharp.

The same Siberian culture continues across North America, or at least Alaska, northern Canada and Greenland, which is logical when it is remembered that the Eskimos coming from Asia were the most recent to arrive on the scene. It is thus natural to find that they have *bone knives*, *snow scrapers*, and *women's knives* similar to those found across the straits.

On the other hand, the original inhabitants of the rest of the Americas escape easy affiliation with the outside and even classification among themselves. California alone, when the Europeans arrived, was populated by fifty different peoples, most of whom spoke mutually exclusive languages. America apparently passed through approximately the same phases as Europe. Thus some Mexican obsidian knives, up until the arrival of the Conquistadores, strangely resemble knives discovered in the caverns of southwest France and northwestern Spain, and the *sacrificial knives* of the Mayas, which were used to tear out the victims' hearts have nothing about them that is peculiarly original.

The form of the American Indian *scalping knife* varies from tribe to tribe. It should be borne in mind, and here many confusions have arisen, that the Indians of North America — anthropologically speaking — existed in an ambiguous state, partly independent, partly under foreign influence. They often borrowed their tools and their weapons from the whites and 'personalized' them as their own. Many knives which are named after tribes (Palute, Shawnee, Seneca, Crow, Comanche, Osage, Hopi) are Indian in name only.

What has been said about America can be applied to the Pacific cultures, as much to Australia as to New Guinea and Polynesia — at least when con-

In the Islamic world it is common to think of a weapon as a man's piece of jewelry. Side arms have a very important place in Islamic art. The two knives with triangular blades (top and lower left) are 'kuttars' or Indian daggers. The handle of the first (18th century) is iron damascened with gold; that of the second is jade. Length: 14-3/8 and 15-7/8 inches. The knives with horse's heads are also Indian, one a 'kandjar' in gilded steel and the other a 'jambiya' in damascened steel with a grey jade handle inlaid with gold, rubies and emeralds. Length: 14-5/8 and 20 inches. The war axe is called a 'tabar': it is of Persian or Indian origin (17th or 18th century). The edge is Damascus steel and is damascened, or inlaid, with gold. The handle is gilded copper, covered in the middle with gold. Length 22 inches. The weapon on the right is a 'kinjal' or 'Cherkess dagger', 18th or 19th century. The blade is steel inlaid with gold. The handle is walrus ivory. Length: 20-1/4 inches. Réunion des musées nationaux, Paris

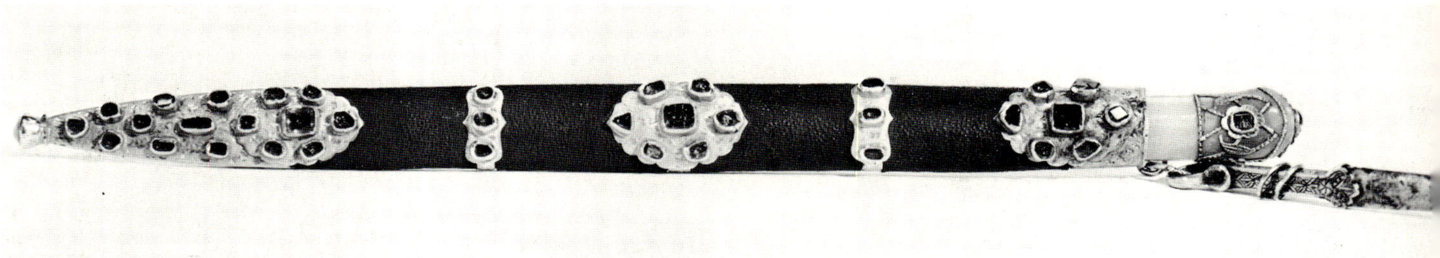

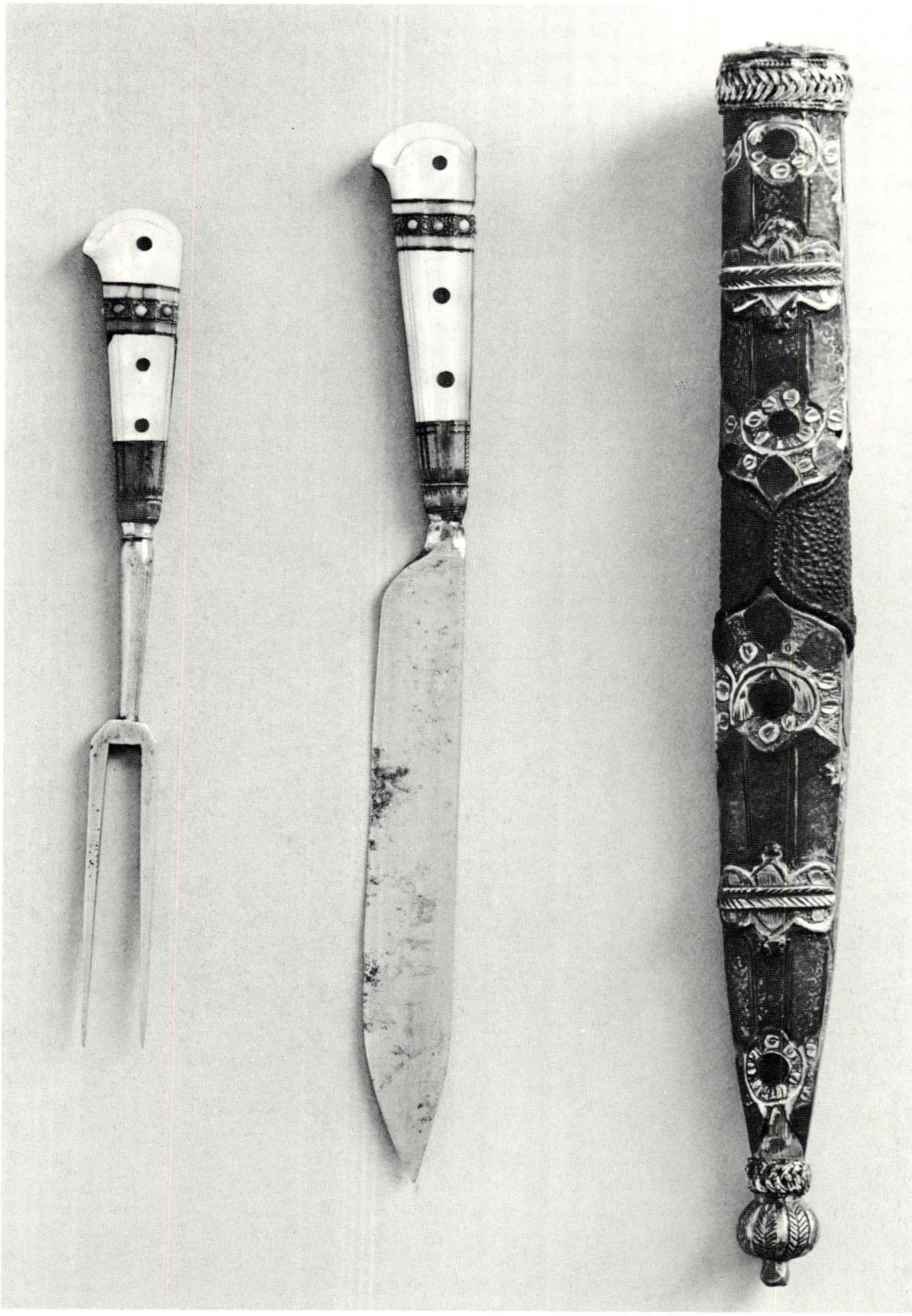

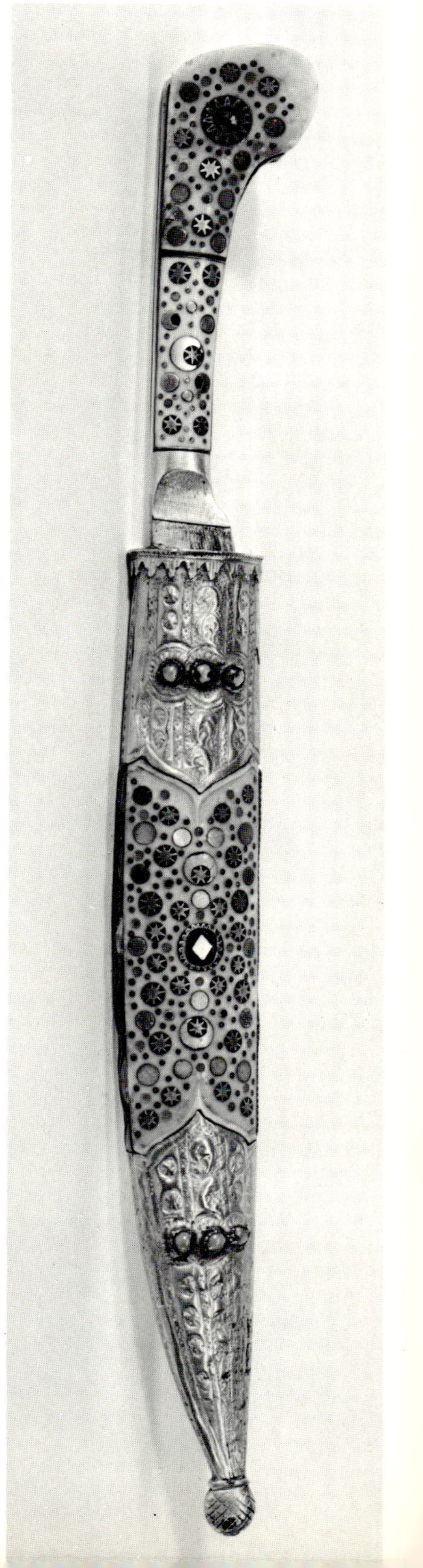

Not all Islamic knives are weapons, however. The pleasures of the table also received their due, as can be seen from the knife and fork on the far left. The double scabbard holds both the fork and the knife. Grimmer functions were at least foreshadowed by a ceremonial weapon, the long knife at the top with ivory hilt and scabbard incrusted with precious stones. Ottoman Empire. 17th century. Schloss Nymphenburg, Munich

Left: This kind of knife can be found in several countries of the Near and Middle East. It could be of Persian origin. The handle and scabbard are incrusted with precious stones. The decoration is of 'astronomic' inspiration. Ottoman, beginning of 17th century. Schloss Nymphenburg, Munich

The handle of the dagger of Selim I, 'the Cruel' is rock crystal inlaid with golden motifs. The finely damascened blade is ornamented on either side with six turquoises and a ruby. Topkapi Museum, Istanbul

sidering what is traditional. Stone is the dominant material for knives although exceptions would have to be made for certain Papuans, who had already been using iron for some time, and for the Fijians who preferred to use fire-hardened bamboo. Delicate operations have even been performed with such knives, which can be as sharp as razors.

The most recent study of African languages notes a total of about a thousand, most of them in the Sahelsudanese zone, and this number must be multiplied by two or three to obtain the number of ethnic groups. So it should be no surprise that there is no common denominator between the knives of the Wolofs of Senegal and those of the Makua of Mozambique.

These peoples are the victims of a cultural loss, since they no longer manufacture knives themselves, but prefer to buy them from Europeans today, having probably earlier acquired them from Arabs and Berbers. On the other hand, the Mandingos still accord the blacksmith an important social position, as is the case with several peoples of the Sudanese belt. Not only does he belong to a particular caste, and not only does he manufacture knives, metallic objects and a host of other instruments, but it is he who performs circumcisions and delivers marriage proposals.

The Fulbe are considered the most skilled blacksmiths of this region: they are found from Senegal to Chad. Félix Dubois (in *La Vie Noire*) describes one of their furnaces in these terms: 'a brazier in the ground fanned by two goatskin bags which end in a metal tube that a helper holds between his legs, pumping them. Such is the bellows of the smithy.'

But what strikes the casual observer, whether he looks at Moroccan or Bateke daggers, Central African *throwing knives* or Mambutu *execution knives*, besides the diversity of shapes, sizes and decorations, is the importance given to the sheath, itself often a work of art.

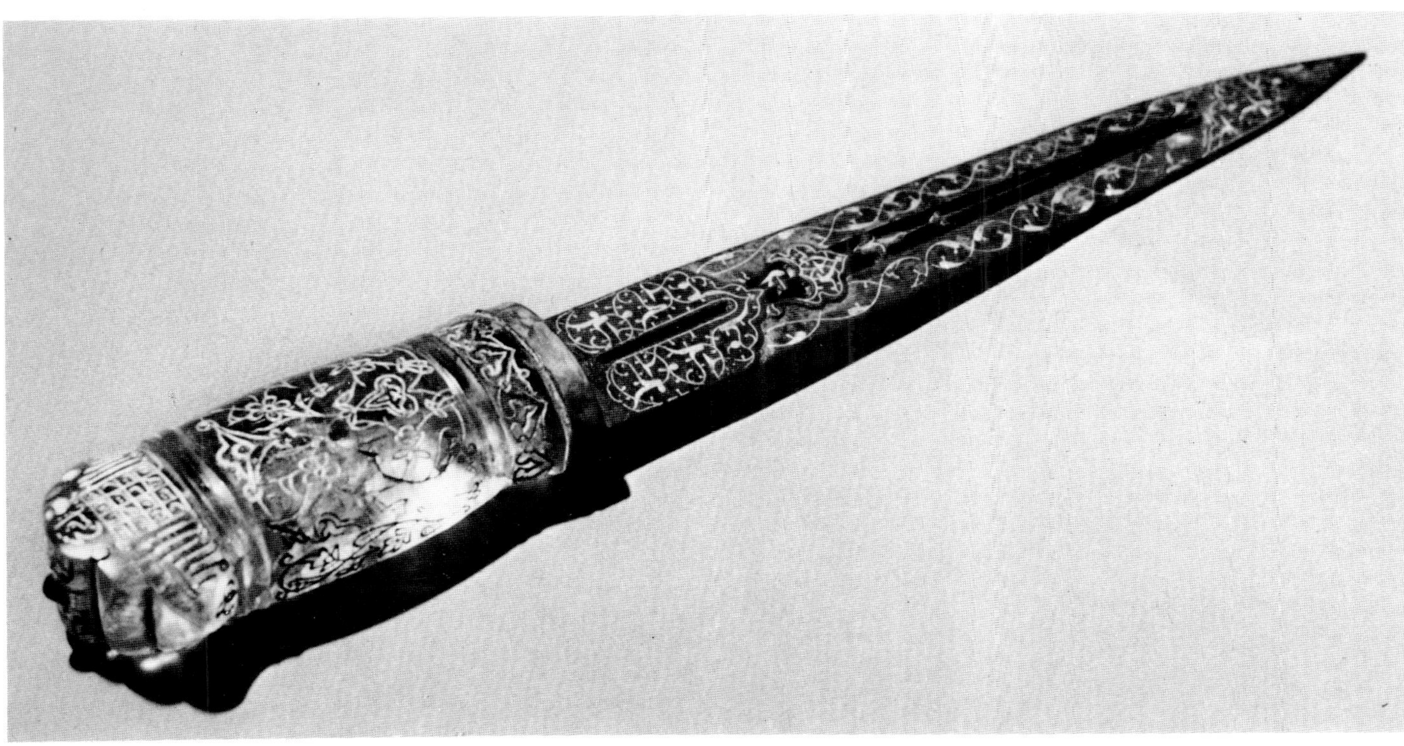

Above: A dagger of steel inlaid with silver and copper. The handle and blade are of one piece. The handle terminates in a lion's head. The blade decorations show a lion hunting. Persia, 18th century. Musée d'ethnographie, Geneva

The dagger (far left) is engraved and nielloed, and has an ivory handle. It clearly shows Islamic influence even though it comes from Kandy, Sri Lanka. 18th or 19th century. Length: 11-1/2 inches. Musée d'ethnographie, Geneva

Left: The Persian knife is in late Safavid style (17th century). The handle is ivory with precious stones. The blade is decorated with arabesques and carries inscriptions. Schloss Nymphenburg, Munich

This is a Syrian dagger dating from around the 10th century. It was certainly used for full-dress ceremonies. Musée du Louvre, Paris

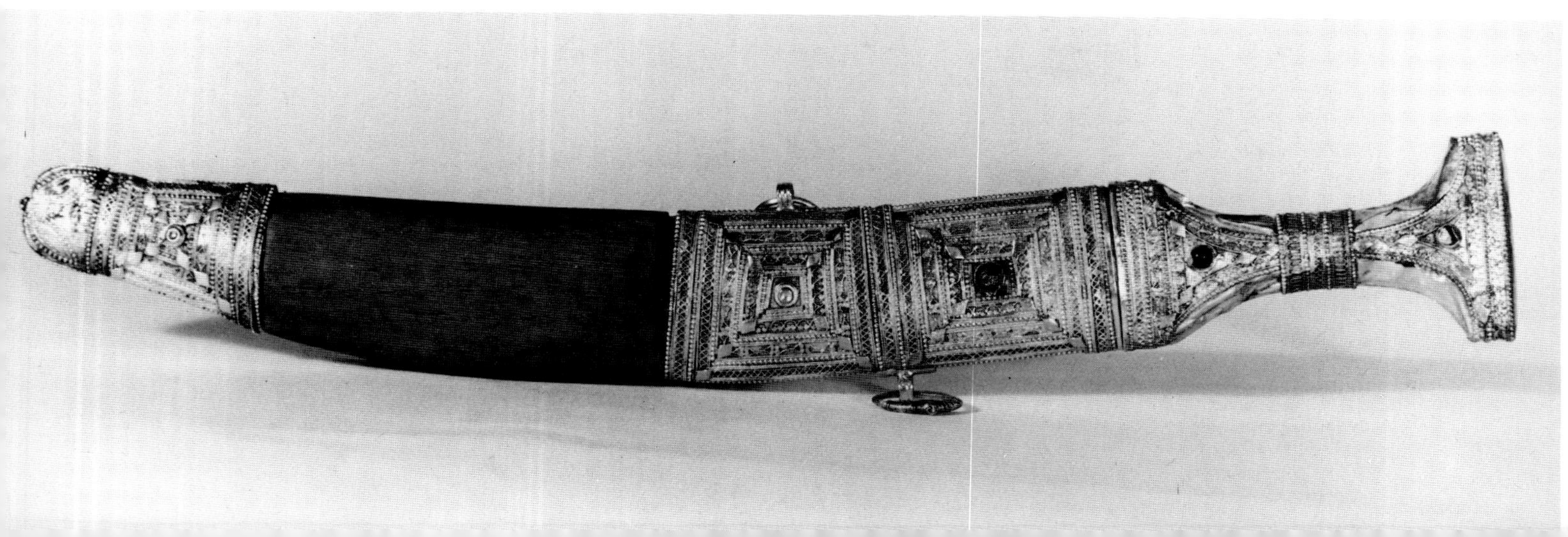

An Indian knife-dagger with a copper handle covered with jade and three rows of pearls. Each row is set off by iron and integrated in a gold and red composition. The end of the handle has a small ring. Length: 12-5/8 inches. City Museum, Sheffield

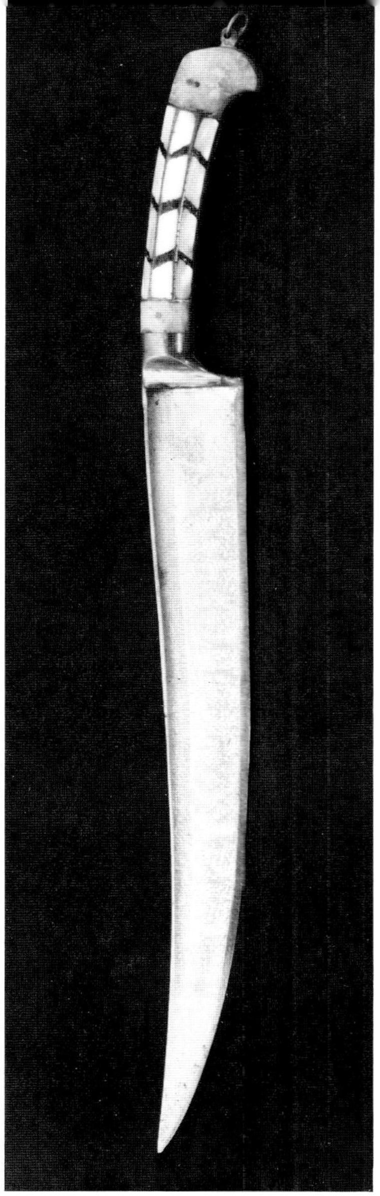

Far right: The 'kadjar' is a dagger dating from the Moghul period. Indian art was at its most magnificent during that epoch. The blade is treated in 'trompe-l'œil'. India, 17th century, Musée du Louvre, Paris

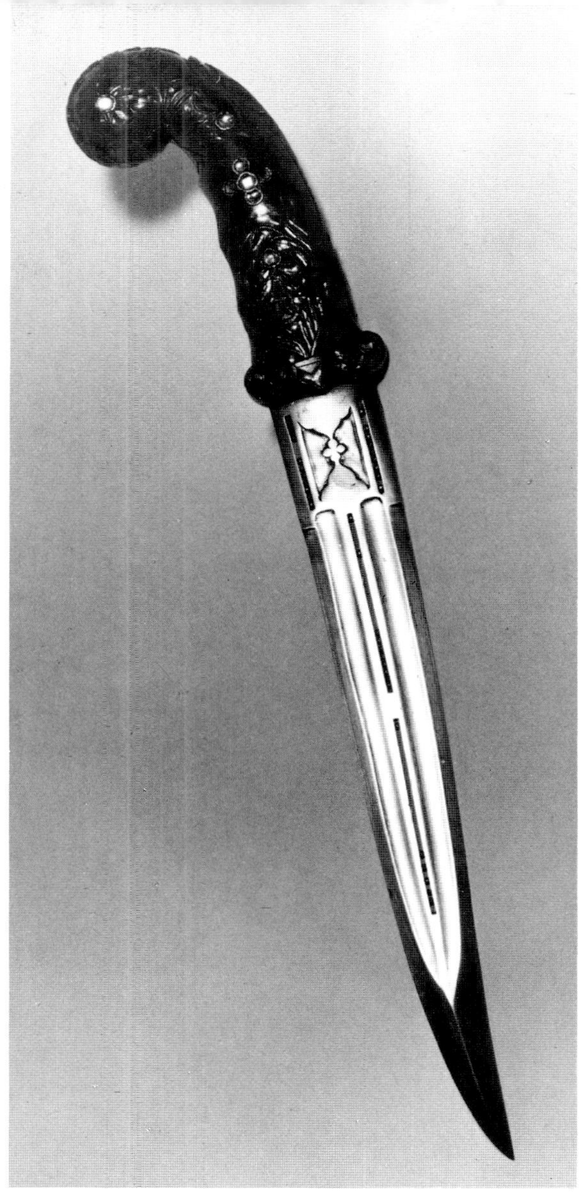

Below: It is difficult to determine with certainty whether this 'kard' is of Indian or Persian origin. The blade is Damascus steel with gold damascene work. The handle, in walrus ivory, is inlaid with gold and rubies. The scabbard is wood covered with rose-colored Indian silk with silver motifs. The chapes are decorated with gold scrolls and rubies. Length: 15 inches. Musée du Louvre, Paris

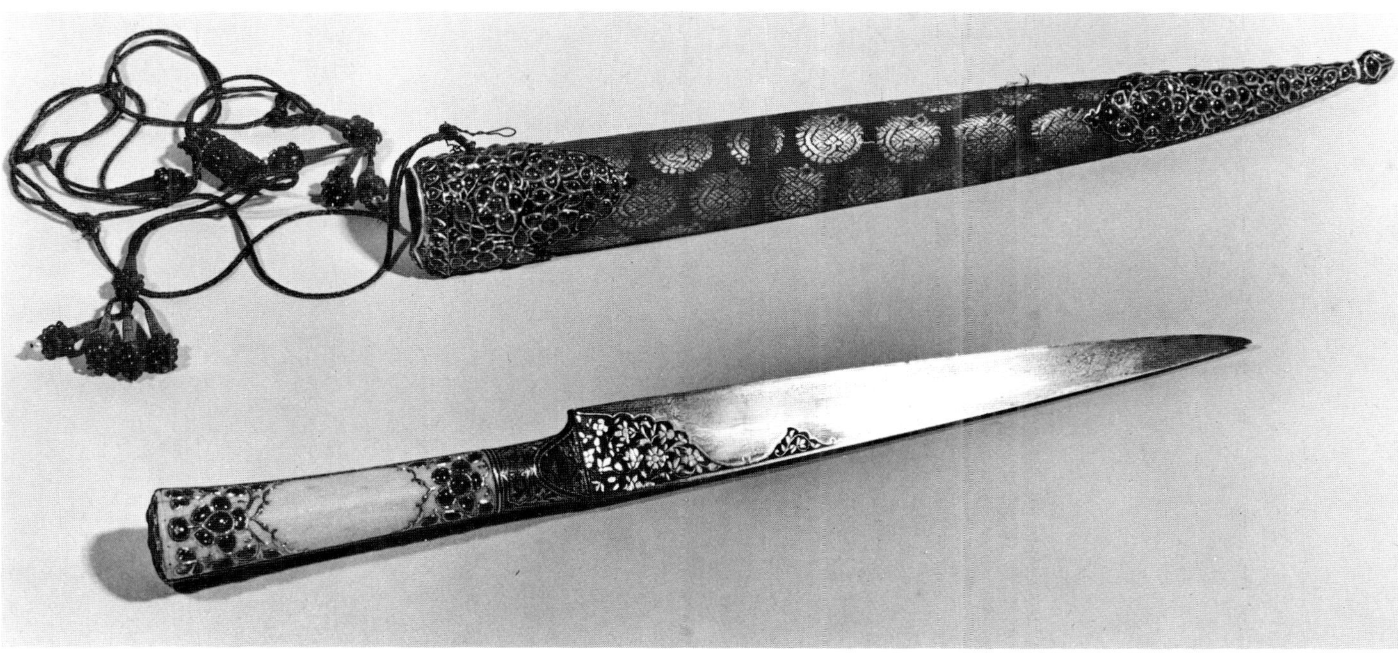

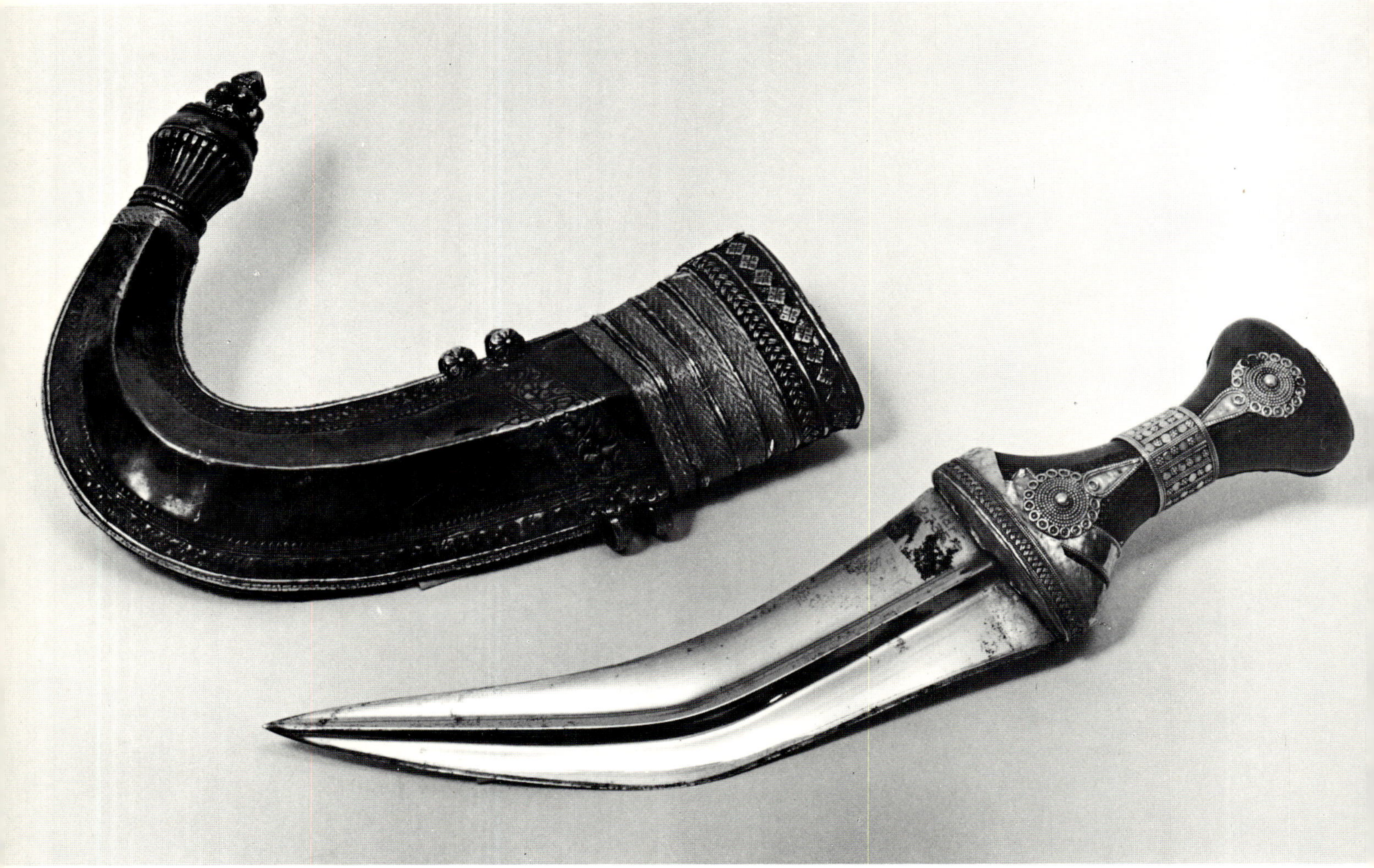

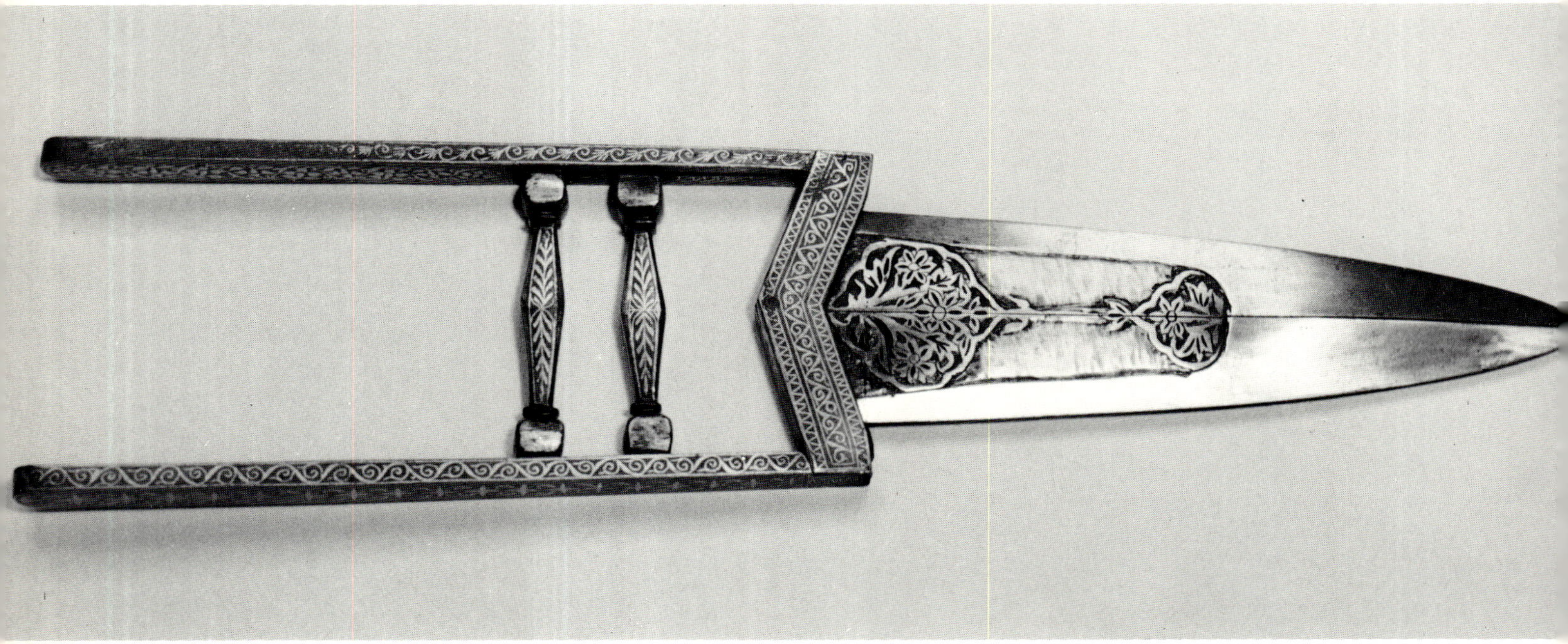

Left: A 'jambiya' from South India. The origin of this type of knife is to be found in the Arabian Peninsula. Afterwards it was transplanted to all the Islamic countries. The materials used are horn, gold, silver and silver-gilt. The blade is of recent manufacture. Musée du Louvre, Paris

Left: Even though it has undergone Islamic influence, the 'kuttar' is native to the Indian sub-continent. In this piece the motifs are abstract, but it often happens that Hindu themes are found in Moghul art. The blade is triangular and the handle is double. The blade opens to reveal a hidden point. West India, 18th or 19th century. Musée d'ethnographie, Geneva

Right: A dagger, in the marvelous Tibetan Buddhist style, is a 'phur-bu' or sacred bronze ghost dagger. Found in the Himalayas, it is used in ceremonies that symbolize the victory of wisdom over the selfish passions. It is held by a masked dancer. The weapon is symbolic: it is surmounted by a 'dorje' which evokes the adamant invincibility of truth. The faces recall the work of wisdom on the different planes of the world. The triple blade cuts desire, hate and illusion, and comes out of the mouth of a sea monster which symbolizes the insatiable desire to save all beings. Tibet. Musée Guimet, Paris

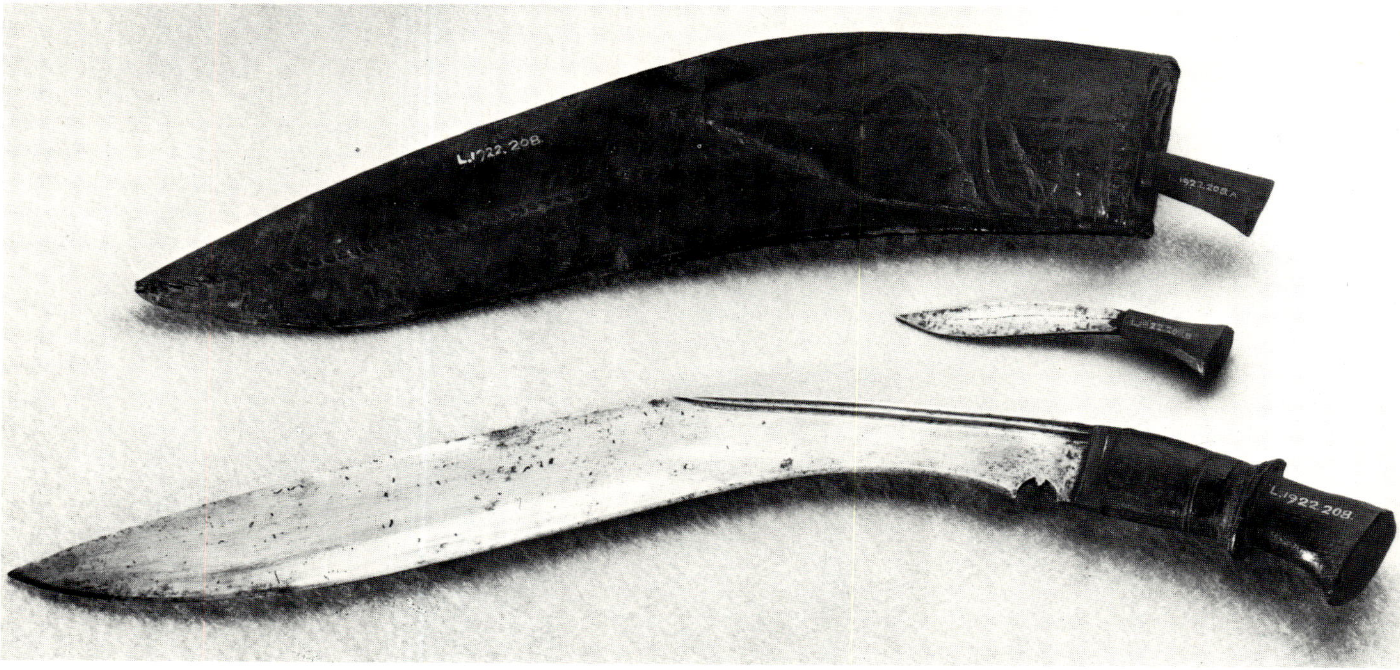

The 'kukri' is the fighting knife of the Gurkhas, the largest ethnic group in Nepal. As such it was used in the British Indian Army. Here the blade is iron, the handle is hardwood. The scabbard, in black leather, has two sheaths for smaller knives. Length about 12 inches. City Museum, Sheffield

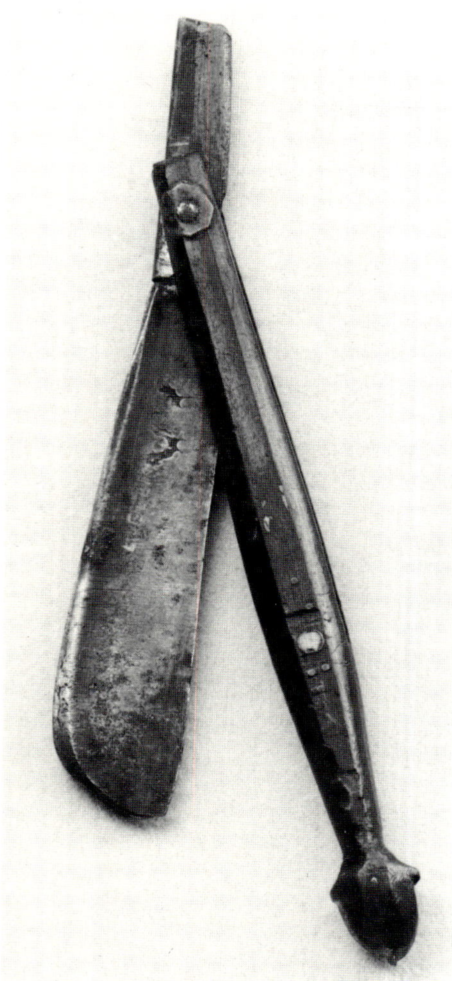

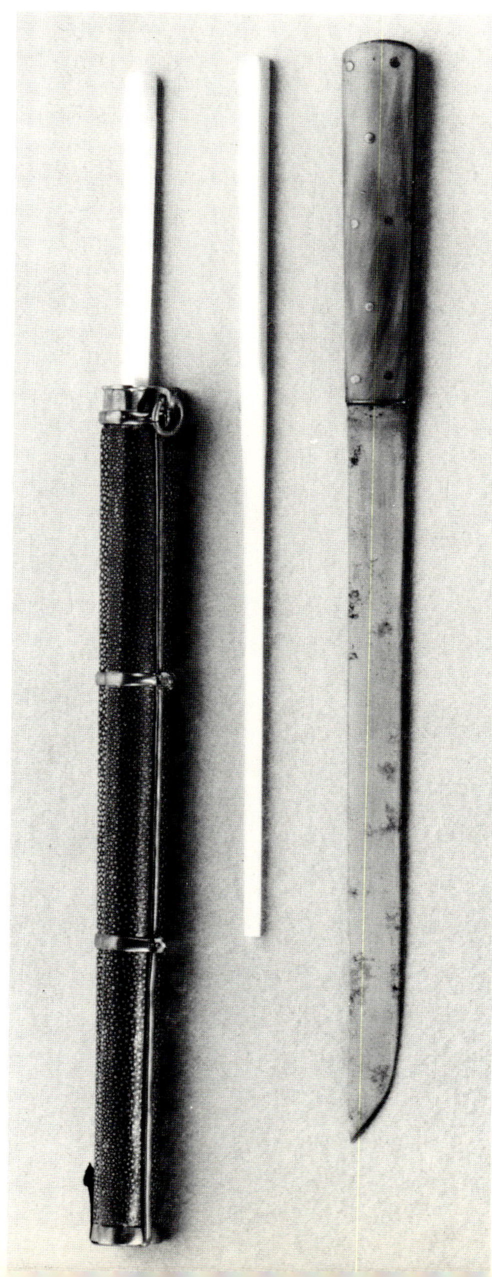

Right: These Japanese daggers or 'tanto' always have the same basic shape, but there are examples which are very simple such as the one on the left, or more elaborate, like the one on the right. The scabbard of the first holds an additional knife called the 'kozuka'. It dates from the Tokugawa Shogunate (17th-19th century). On the right, the handle and scabbard of carved ivory are decorated with scenes of daily life. Period unknown. Both knives are about 12 inches long. Musée d'ethnographie, Geneva, and City Museum, Sheffield

Far left: A Chinese razor, which is apparently of European influence, has a handle of wood with brass and ivory inlay. Length closed 6-1/2 inches. City Museum, Sheffield

Left: A bamboo scabbard covered with sharkskin which contains a table service consisting of a knife with an ivory handle and two ivory chopsticks. Length of the knife: 12-1/2 inches. China, 19th century. City Museum, Sheffield

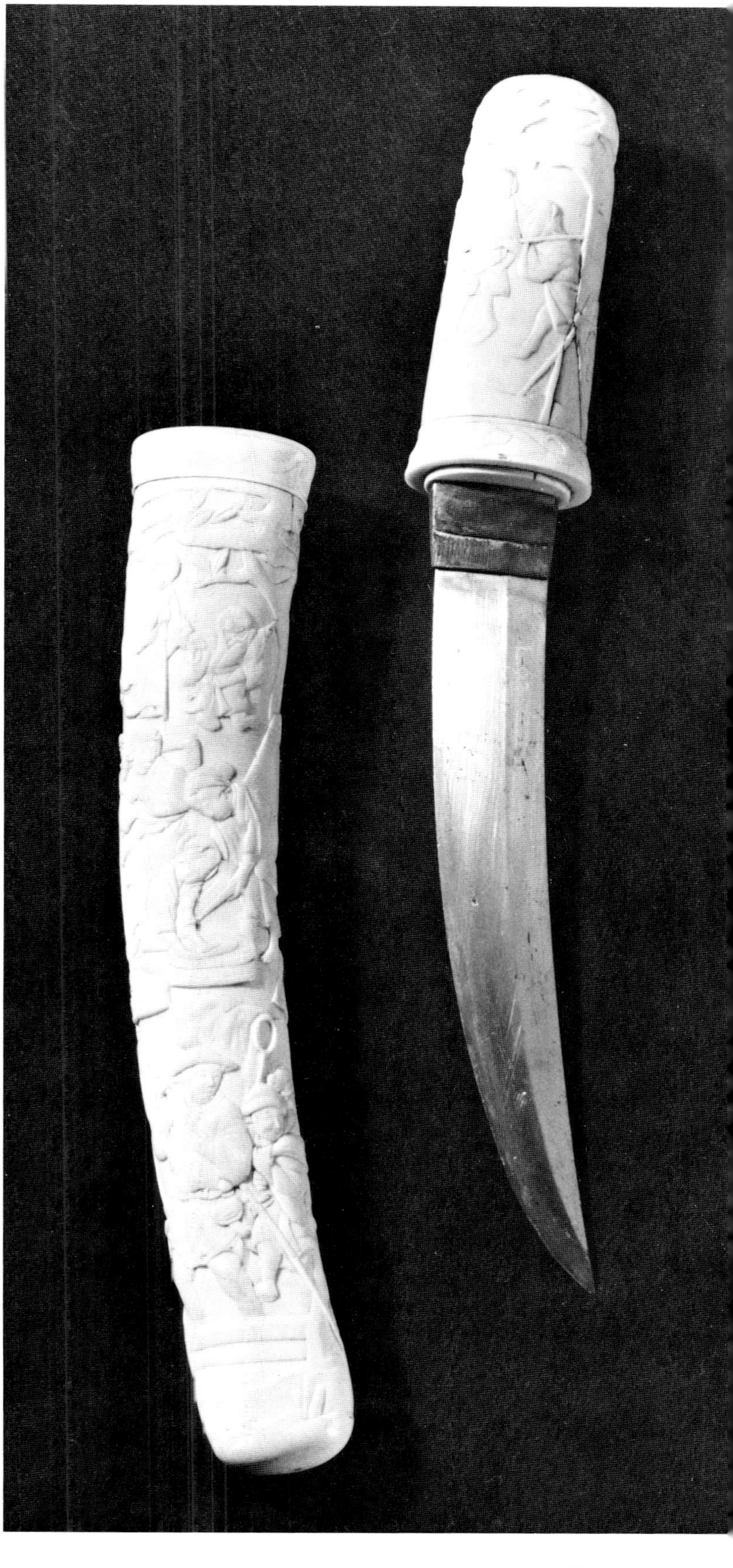

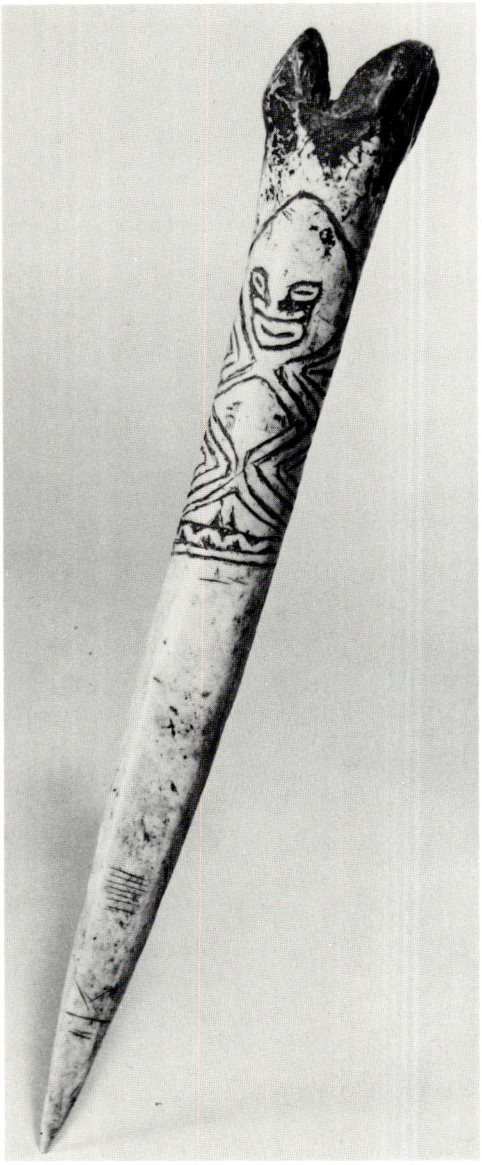

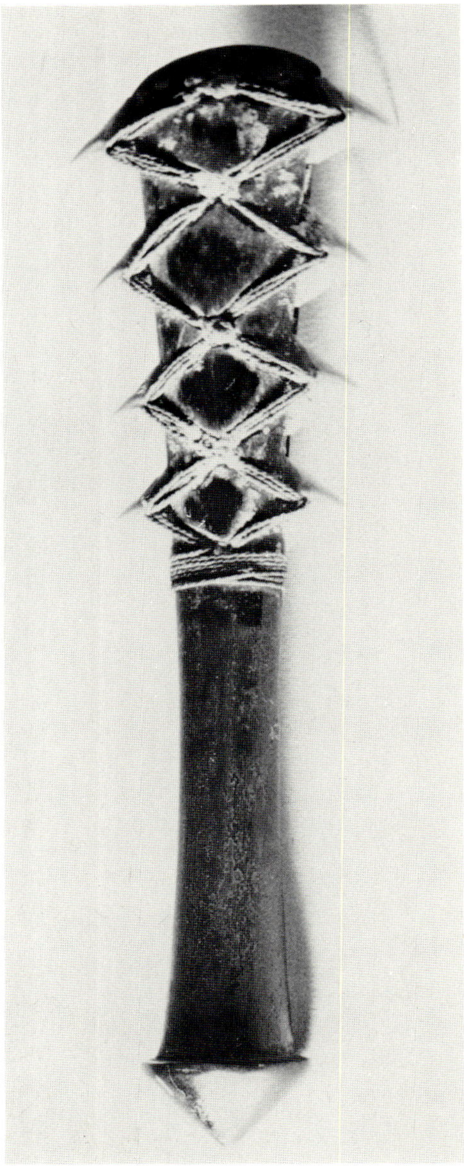

Far left: New Guinea is the second largest island in the world. The interior has not been completely explored and some of the inhabitants live in the stone age. They also use bone, as can be seen from the dagger on the left made from a cassowary femur, and bamboo as seen in the knife below, which is a head hunter's weapon. Musée d'ethnographie, Geneva

Left: This weapon, made of wood and shark's teeth, is a woman's dagger from the Matty Islands, which are a Micronesian culture although they are located in Melanesia. Musée d'ethnographie, Geneva

Despite the edge shape, this knife is not a Malayan kriss, but a Tibetan or Nepalese ritual moulded-iron razor. 20th century. 8-1/2 inches. Musée d'ethnographie, Geneva

Left: The aborigines of Australia come from one of the most 'primitive' ethnic groups in the world. The Aranda or Arunta in the center of the continent are one of the least known tribes. Not so long ago they still used stone knives like this. 7-1/2 inches. Musée d'ethnographie, Geneva

Below: The Eskimos cover a vast area from eastern Siberia to Greenland. Their culture differs from one region to another, as can be seen from the knives below: the first has an iron blade; the bone scabbard is decorated with pictures of caribou (Alaskan Eskimos). The woman's knife (Ooloo), below right, of black schist, belongs to other Alaskan Eskimos (Point Hope); it is used to scrape sealskins. The bone 'snow knife' — or ice scraper (below, left), is the work of Greenland Eskimos. Musée d'ethnographie, Geneva

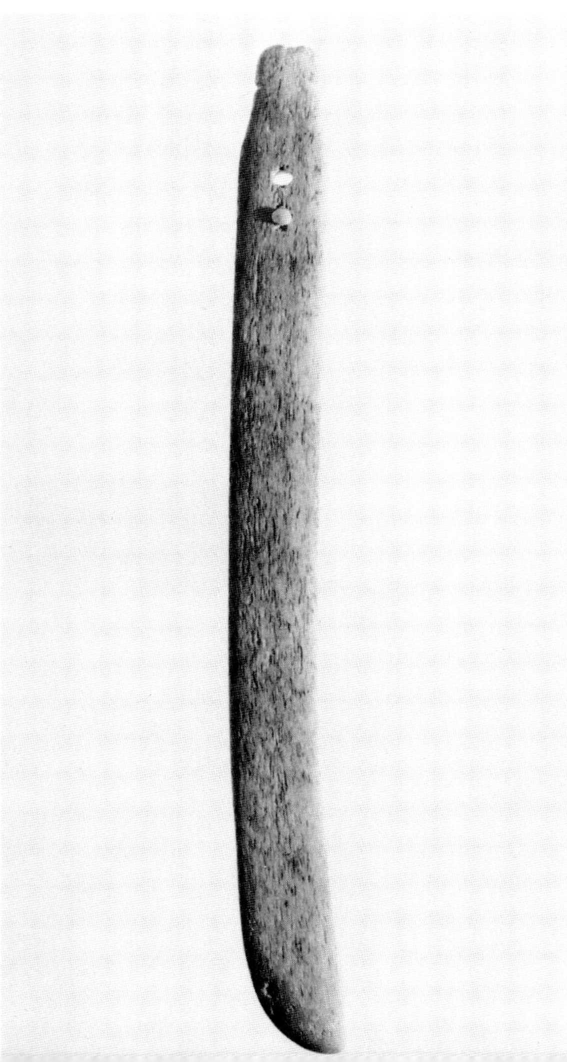

This book started with the stone age and now ends with the stone age with this obsidian blade from pre-Columbian Mexico. Obsidian was the favorite material for ritual knives. Note the Solutrean-like leaf work. 4 inches. Musée d'ethnographie, Geneva

Glossary-Index

This glossary contains various names of knives, cutlery and related terms. It is by no means exhaustive; rather, an attempt has been made to define or develop some words not always fully defined in the text.

For other words, the reader is referred to the page where they appear or are illustrated, or sometimes, to another word in the glossary indicated by an asterisk*. It should be borne in mind that not all cutlery terms are precisely fixed. The same term might be used to refer to different things, and there is variation in the way different cutlers use the same term.

Knife ends are made in various shapes, some pointed, others rounded off with one or two cutting edges. The word point is generally kept for the sharp ends. Here are a few types: 1. scimitar 2. bill-hook or pruning knife 3. skinning knife 4. sabre 5. clip 6. spey 7. spear 8. needle 9. round 10. coping 11. sheep foot (cutting edge right) 12. Turkish 13. yatagan type.

acinaces (Latin, from Greek *akinakes*, of Persian origin) Short, stright dagger as used by the Persians, Medes and Scythians.

all-purpose knife See p. 77

annealing See p. 12. Heating of a metal to a temperature not higher than that of a physico-structural stability followed by slow cooling. This process augments the ductility of a metal and decreases its resistance to deformation.

antenna knife See p. 123

Arkansas toothpick See p. 62

arming dagger (arming sword) A very short medieval sword worn on the side. Any cutting or thrusting weapon used in the Middle Ages.

astronaut's knife See p. 125

automatic knife See p. 100

bacon knives See p. 117

baker's knife See p. 118

bastardeau A small knife used at table or as a weapon, carried in a special pocket of a sword or dagger scabbard.

bayonet See p. 31

bayonet-sword See p. 31

becquoysel See p. 123

bee keeper's knife A knife used for removing the wax caps of honey cells. Its shape is like a planer and it is not properly speaking a knife, but there are hand models.

belt dagger See p. 139

belt knife See p. 72

bendo See p. 145

blade section See edge*

blade See p. 21 and illustration p. 165

blank The bar of metal from which a blade is made whether the method is stock removal or forging.

blast furnace One or a group of cylindrical constructions meant for melting and reducing mineral ore to obtain the metal. At the base, the melted carbonized iron alloy is gathered.

bleach (verb) 1. To clean blades and sometimes the handles. 2. To refine a casting in order to transform it into steel. 3. To clean a metal by reheating it and then quenching it in water mixed with acid.

blood groove On some hunting knives a groove near the spine of the blade and parallel to it.

bloodletting knife See p. 121

blunted Certain kinds of 'points' of cutting instruments made inoperative by a 'button', of steel, leather or other material.

bolo A large, single-edged knife used in the Philippines as a weapon or cutting tool, often resembling a machete. The U.S. Army once issued bolos for combat. See p. 42

bolster 1. A band of metal on kitchen knives and some other knives at the

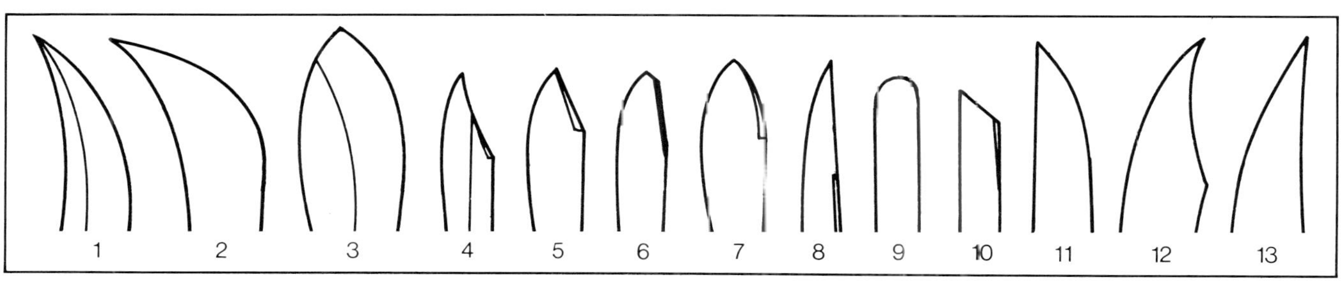

point where the blade and handle meet. 2. A mounting on either or both ends of a folding knife.

boning knives See p. 117

boot dagger See p. 66

boot dirk See p. 66

boot knife See p. 66

Bowie See p. 40

Bowie bayonet See p. 42

Brackmard A short sword with a broad and sharp blade and with two edges, used during the 13th to 15th centuries.

brass An alloy of copper and zinc, sometimes called 'yellow copper'. Used in cutlery mainly for the trimmings of handles.

breach knife See knife*

bread knife See p. 72

bride's knife See p. 131

budding knife Small knife for grafting buds.

butcher's knife See p. 117

butcher's steel A roughened steel rod on which to sharpen knives.

butterfly knife See p. 100

button An ornament at the end of a handle.

cabbage cutter Short sabre carried by French infantry from 1831 to 1866. Synonym of cut-cut*.

camper's knife See p. 44

capuchin knife See p. 95

carton knife A kind of scraper with a long handle to cut cardboard.

carving knife A kitchen knife; compared with the chef's knife, the blade is not broader than the handle. It is used for carving poultry, fowl, and slicing middle sized joints of meat. See p. 73

case See p. 27

cataract knife See surgical knives*

cauterizing knife An instrument used to disinfect wounds.

cementation To heat intensely a piece of metal in contact with a metallurgic cement which combines chemically to produce a new product.

chain mail piercer A strong dagger of the 14th-15th centuries, with a reinforced point used to pierce hauberks and chain mail armors.

chape A mounting or fitting on either end of a scabbard or sheath, particularly the point end.

Châtelperron point See p. 9

cheese knife See p. 77

chef's knife The largest of kitchen knives; used mostly for cutting and chopping large pieces of meat. See p. 73

Cheseldon knife See surgical knives

choil A cut out part of the blade between the edge and the hilt.

chromium A white metal with similarities to iron and manganese. Often used as a covering for steel or its alloys because of its resistance to atmospheric and chemical corrosion.

cinquedea See p. 32

circumcision knife A ceremonial knife used for removing the prepuce. For a long time the Jews used circumcision knives made out of stone.

clam knife Its shape can vary but it always has a bolster and is double-edged. The present professional clam knife is rather narrow, has a thick blade and is long and sharp. The oyster knife is similar.

clip A portion of the spine of a knife that drops away from a straight line to meet the point.

cobbler's paring knife See p. 120

cod fisherman's knife See p. 121

coin knife See p. 131

commando dagger See p. 38

convent table knife See p. 71

corn-cutter An instrument with a slightly curved blade used for cutting corns and callouses.

counter-guard In the 16th and 17th centuries, the rings and flourishes around the handgrip of a sword meant to protect the hand.

courtly arms See p. 31

couteau à la d'Estaing See p. 97

coutelet (cutlet) Long ago, a small knife for paring fingernails.

coutelle Short medieval sword. The coutelle was distinct from the arming knife because it had two edges.

covers Plates of horn, shell, ivory, plastic, wood, etc. which cover the inner metallic plate of a folding knife.

craft method (bench method) See pp. 75 and 76

crooked knife See p. 96

cross-belt (shoulder-belt) A leather belt worn across the shoulder to hold a sword or dagger at the waist.

cup guard See p. 34

curved knife See saddler's knife*

custile (*costile*) According to some, it

is a sword with a broad blade and a very sharp point; according to others it is a shaft weapon.

cutler's marks See p. 22 and following pages

cutlass 1. Originally a large knife meant for various uses: the hunt (for killing the animal as well as dressing it), war, etc. 2. A short sword with a single edge for naval use. 3. A large kitchen knife. 4. In the 16th century it was sometimes synonymous with a scimitar (which is any very curved weapon with a large back and a single edge.)

cut-cut 1. A brush clearing knife. 2. Name of various long blade knives

Some different types of folding knives include: 1. electrician's 2. alpine 3. shepherd's 4. trapper's 5. Belgian 6. linoleum 7. twin-bladed trapper's 8. Boy scout 9. big-game 10. rancher's 11. sheep-foot 12. fisherman's 13. German 14. "Tinker"-Swiss Army 15. handyman's 16. fisherman's-Swiss Army 17. mariner's.

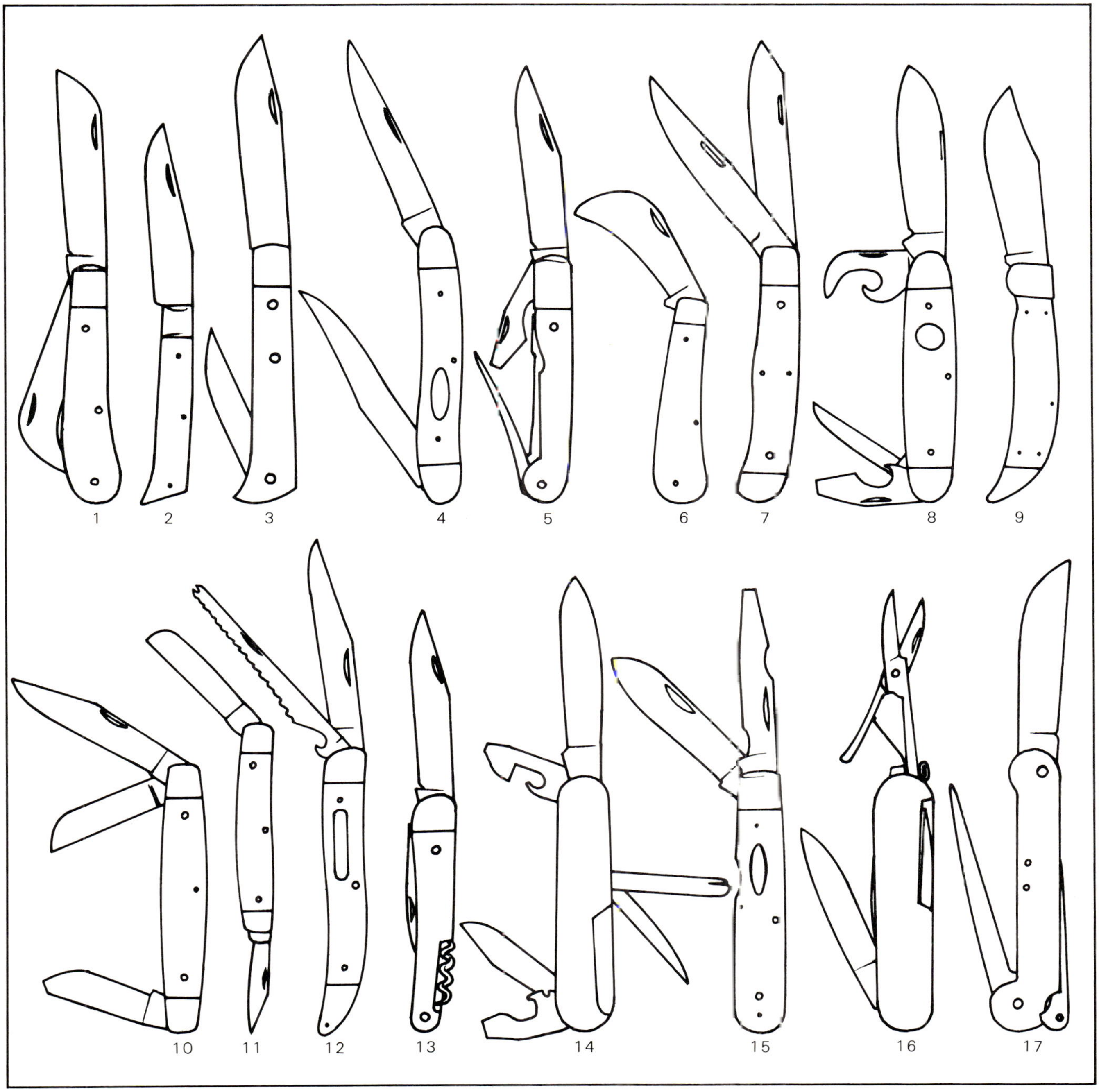

found in West Africa and Indochina, i.e. former French colonies.

cutlet Long ago, a small knife for paring fingernails.

cutting needle See p. 118

dagasse A dagger with two edges and a wide blade, similar to a cinquedea*.

dagger The least well defined of short weapons. One dictionary defines it as a 'short sword with a narrow and sharp blade.' Another says it has a wide blade. Dagger has also been used as a term for weapons with a single edge, even hunting knives. Finally, the term is used by bookbinders to mean an iron blade with handles at both ends used for scraping skins.

daguette A small dagger.

dah See p. 144

Damascus and *damascene* The two words are often confused. Damascus properly refers to the steel (which is also called 'Indian steel' or 'Wootz steel') obtained by hammering hot spongy iron without attaining the temperature of fusion. The surface becomes steel irregularly by contact with the carbon of the heating matter. The bar is tempered and the irregular mixture of steel and iron produces moiré patterns. Damascening refers, properly speaking, to the inlay of gold or silver wire on an iron surface by hammering. Square or twisted wire is first chased into the metal and then hammered. The surface is then polished. Thus a blade can be both 'Damascus' steel and 'damascened' at the same time.

decorating knife Name given to refer to various knives used to 'sculpt' food. See p. 81

defensive weapons See p. 31

derava Arab dagger attached under the left arm by a belt.

dolphin fisherman's knife See p. 20

double-edged See p. 9

driver's knife See p. 104

drop-press See p. 75

drop-stamp A hammer moved by a wheel with cams used in the forging of knives.

'*dull knife*' A knife with a dull edge used by curriers for removing the hair of animal skins.

eagle's head knife See p. 96

ear-curette See p. 97

edge The cutting part of a blade. See p. 21

edge-tool making Branch of the hardware industry which specializes in the manufacture of iron with cutting edges used by farmers and certain crafts: axes, scythes, plough shares, various butchering instruments, etc.

electrician's knife See folding knives*

electroplating A process which electrically produces an overlay or coating of gold, silver, tin, etc. on another metal.

English knife A cutting instrument used by farriers for paring hooves.

escutcheon plate A small plate stamped with a sign, coat of arms etc. soldered to the bolster of a table knife or fixed elsewhere. Also found on other kinds of knives.

estoc 1. A sword of the 15th and 16th centuries with a strong blade and a sharp point for thrusting. 2. The point of a side arm. Fr. *D'estoc et de taille*: to cut and thrust.

estocade See p. 32

eustache See p. 95

execution knife Its form varies according to societies and whether the victim was stabbed or had his throat slit. They are found all over the world.

exotic knives See p. 141 and following pages

exposition knives See p. 131 and following pages

false edge A sharpened edge on the back of a blade.

factory production See p. 76

fauchard A shaft weapon with an iron in the form of a knife with a single blade. It sometimes has a hook at the end.

fauchon A cutlass with a single edge, a strong blade which flares out towards the end and sometimes terminates in a point.

fighting utility knives See p. 42

fillet knife See p. 81

firearms See p. 31

fisherman's knife See p. 121

fish knife A knife in the shape of a spatula for opening and serving fish at table. See also slice*

fixed blade knife A non-folding knife.

flange A round and moulded part of the blade near the handle which often replaced the bolster until the end of the 18th century.

flat knife A knife with two edges and no point. See p. 75

flaying knife See p. 117

flissa A yatagan used in Kabylia with a scabbard made of two pieces of wood held together by steel bands.

folding bayonet-knife See p. 96

folding knives See p. 95 and following pages as well as illustration p. 161

food sculpting A kind of carving used to give food (generally vegetable or fruits) a decorative look.

foot knife See knife*

French knives See p. 75

friend's knife See p. 123

frozen-food knife See p. 76

fruit knife See p. 75 and following pages

fulcrum blades See p. 75. A table knife with protruding base or bolster that keeps the blade from touching and soiling the table cloth when placed down. Came into use around 1827.

furgeoir Small medieval knife for picking the teeth.

furrier's knife See p. 121

gardener's knife See p. 121

gilder's knife See p. 118

golok See p. 145

grafting knife See p. 121

grapefruit knife See p. 73

grips Plates of horn, shell, plastic, ivory, wood, etc. which constitute the handle of a fixed blade knife and which cover the tang.

groove knife See p. 96

guard A metal band placed between the blade and the handle; its edges protrude beyond the handle vertically to keep the handle from sliding down to the blade.

hachereau French word used for a hachet used to cut the joints of game when a game dressing knife is not enough.

hachet-dagger See p. 141

ham knife See p. 117 and illustration p. 73

hand arms See p. 30

handle 1. See p. 21. 2. In folding knives the handle also means the case made up of plates and covers, not counting the blades and accessories.

hansart French name of several different butcher's instruments but mainly a kind of cleaver.

herring fisherman's knife See p. 121

hilt or *ring* A band of metal or thickened part where the blade meets the handle, with or without quillions. By extension the complete handle itself.

hog butcher's knife See sausage knife*

hollow grind A bevelling of a blade that is concave on both sides.

hollow knife See p. 125

honorific arms See p. 31

hook knife A medieval knife (or short sword) with a hook and an extremely narrow blade, which slipped in between joints of armor.

household knife See p. 78

hunting bayonet See p. 31

hunting knife See p. 29 and following pages

interbone knife See surgical instruments*

ice scraper See p. 157

jambiya See p. 152

jambette See p. 95

kalifa See p. 145

kandjar An Indian (Mahratti) dagger with a double-edged and double curved blade. The pommel takes on the shape of an axe-iron.

kará See p. 151

katar See kuttar*

katchik See p. 145

kerneler See p. 123

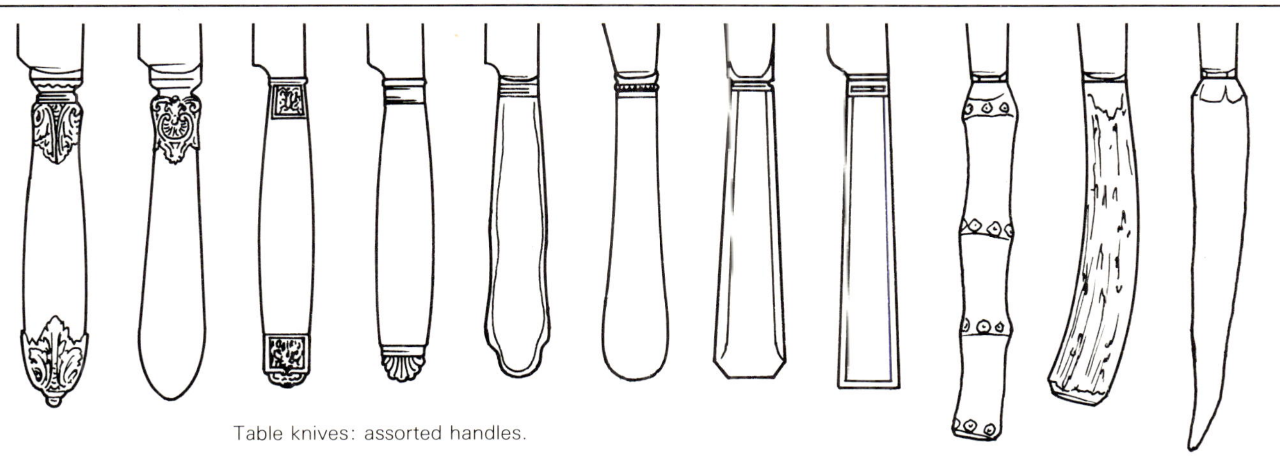

Table knives: assorted handles.

khodmi See p. 142

kick A short tooth on the heel of the blades of folding knives which keeps the edge from touching the spring.

kindjal See p. 146

kitchen knives See p. 71 and following pages

knife In the widest sense, any edged cutting instrument with or without a handle, manual, mechanical or electric. (In this sense one can speak of a plough's 'knife'). In the narrower sense, 'knife' usually designates short bladed instruments, but custom has defined their dimensions. Some authors only speak of knives as those with one edge, but this is not common usage. Generally speaking, objects which have the blade perpendicular to the handle or mounted on a shaft are not knives.

Yet, a cheese knife has two handles, a harness-maker's foot knife has a half-moon shaped blade and a 'breach knife' was mounted on a pike.

Not all knives cut. The 'heat knife', a wooden instrument used to remove perspiration off the skin of horses is blunt, as is the artist's palette knife.

knife-dagger A name used for various combat knives.

knife for gathering dates See p. 142

knife with advertising See p. 122

knife with a rosette See p. 96

knife with a ruler See p. 133

knife with a safety catch A folding knife with a device, usually a spring loaded tooth that fits into a notch, that prevents the blade from closing partially or by accident.

knuckle-bow See pp. 34 and 42

kozuka Japanese knife which was carried in the scabbard of the samurai sword with another knife and the long sword.

krid or *kriss* See p. 145

kukri See p. 145

kuttar See p. 141

lading See p. 145

Laguiole A small village in Provence, France, where, in a tradition going back many years, very well-made folding knives are manufactured. The knives which come from this village often have a bee engraved on the free end of the spring, and the covers are decorated with a dancer's leg.

lancet 1. A small surgical instrument used for bloodletting, lancing or piercing small abcesses, and for vaccinating. 2. Small knife used by a butcher.

left-hander See p. 32

lenticular knife See surgical instruments*

levantine (dagger) At the end of the 15th and beginning of the 16th centuries it was a dagger whose pommel looked like two 'ears'. The handle was held with four fingers and the thumb pressed between the two 'ears' to give force to the blow.

lithotomy knife See surgical instruments*

liturgical knives See p. 123

long blunt knives See p. 118

machete Long knife widely used in the tropics, with a thick blade for cutting underbrush. It was at the same time a tool and a weapon.

machoera See p. 16

manicure knife See p. 138

Micarta A patented resin usually impregnated into cotton cloth but also other materials often used for handles.

milling To machine cold metal by using a milling cutter, a power tool with blades which turns around an axis.

miniature knives See p. 125 and following pages

miséricorde A dagger with a blade reinforced at the point to pierce chain mail armor. It was used to kill an adversary, or give him the coup-de-grâce — hence its name.

missile weapons See p. 31

molybdenum A white, hard metal akin to chromium and tungsten. In certain special steels it is alloyed with nickel, chromium, and vanadium to increase hardness, tensile strength, facility for thermic treatment and to decrease brittleness. Molybdenum steel offers all the qualities of stainless steels with a greater hardness.

multi-blade knives See p. 97 and following pages

navaja A long knife with a tapered sharp and lightly curved blade used in Iberian and Latin-American countries.

nickel A shining white metal akin to iron. Used primarily in alloys with iron, chromium, molybdenum, copper, etc. Steels with a small amount of nickel (18% chromium and 8% nickel) are used for their resistance to atmospheric, marine, or chemical corrosion. The nickel alloys also have great tensile strength.

nickel-silver This is an alloy of copper (55-70%), zinc, and nickel. Pale yellow or white silver in color, it is primarily used for making the trimmings of knife handles.

nut opener See p. 123

Opinel See p. 114

ooloo See p. 157

outdoor knives See p. 40 and following pages

painter's knife See knife*

palmette A decoration in the shape of a palm chiselled into the bottom of the handle of a knife.

paper-cutter See p. 123

paring knife 1. A tool of the bookbinder, glove maker and leather worker, with a flat blade with curved cutting edge, used to even thickness of skins (skiver). 2. Name of various kitchen and butcher's knives used for trimming meat or vegetables.

pas d'ane See p. 32

pastry cook's knife See p. 118

peeler 1. A peeling knife. 2. An instrument used by a basket maker.

peeling knife Name of various vegetable knives, some of which have a slit in the middle. See p. 72

peg Metal rod riveted to one or both ends of a folding knife as an axis for the blades.

penknife See p. 100

phur-bu See p. 143

pisow-raut See p. 145

pistol-knife See p. 123

plane Name of several carpenter's instruments; some with two handles, (jack plane, spokeshave) etc.

plate-knife A thin narrow knife from the Middle Ages made for slipping in between plates of body armor.

presentation dagger See p. 62

presenting knife A large long knife of the Middle Ages and the Renaissance used for serving meat.

production line See p. 76

professional kitchen knives See p. 71 and following pages

professional knives See p. 117 and following pages

pocket knives See p. 95 and following pages

point See p. 21

pommel A small ball-shaped ornament at the end of a handle of a sword, sabre or dagger.

poniard A hand weapon with a pointed and short blade. It can be straight or curved, with one or two edges. Some long poniards are indistinguishable from daggers.

powder knife A knife used for removing face powder. It was an object of small size, often decorated, and sometimes with a golden blade.

power-hammer A large hammer in a forge used to shape and draw out metal. There are several systems to provide the power: steam, compressed air, etc.

push dagger See p. 43

puukko See p. 141

quillions In a guard of a dagger or sword, each of the branches of the cross. The quillions can be straight, curved, or form an S.

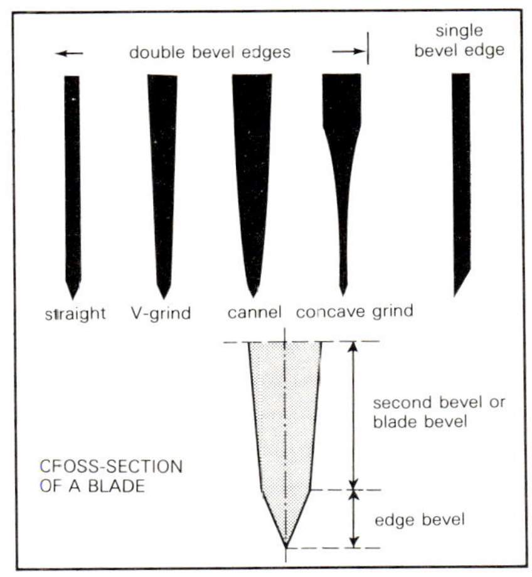

Some peculiar types of blade shapes include: 1. curved in with a guthook 2. Hollow guard 3 and 4. skinning blades 5. skinning blade with almost symmetric cutting edges 6. wide skinner 7. non-symmetric throwing blade 8. shark's tooth design.

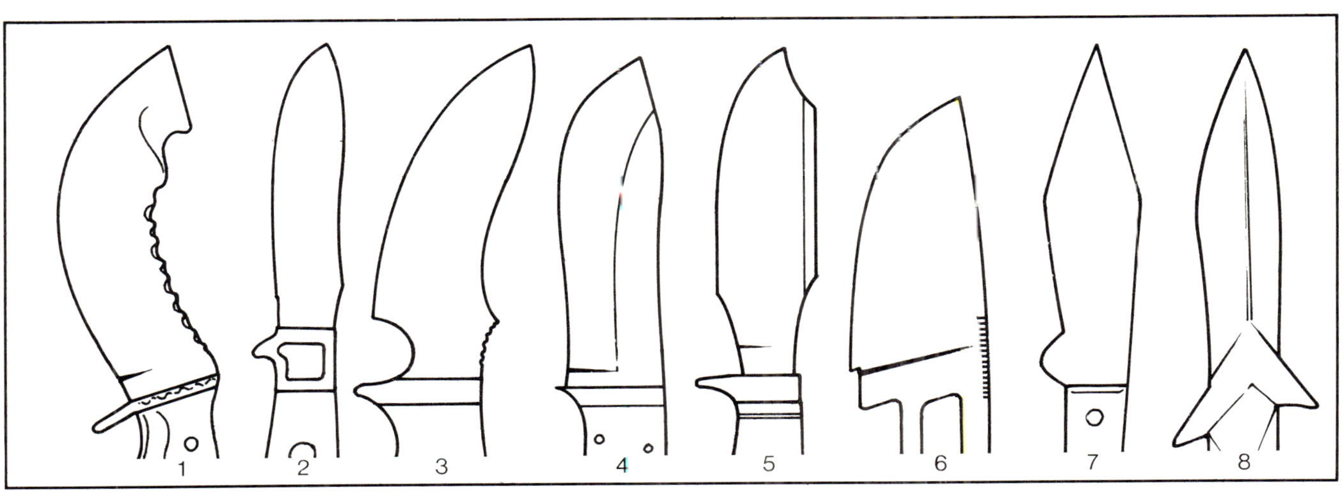

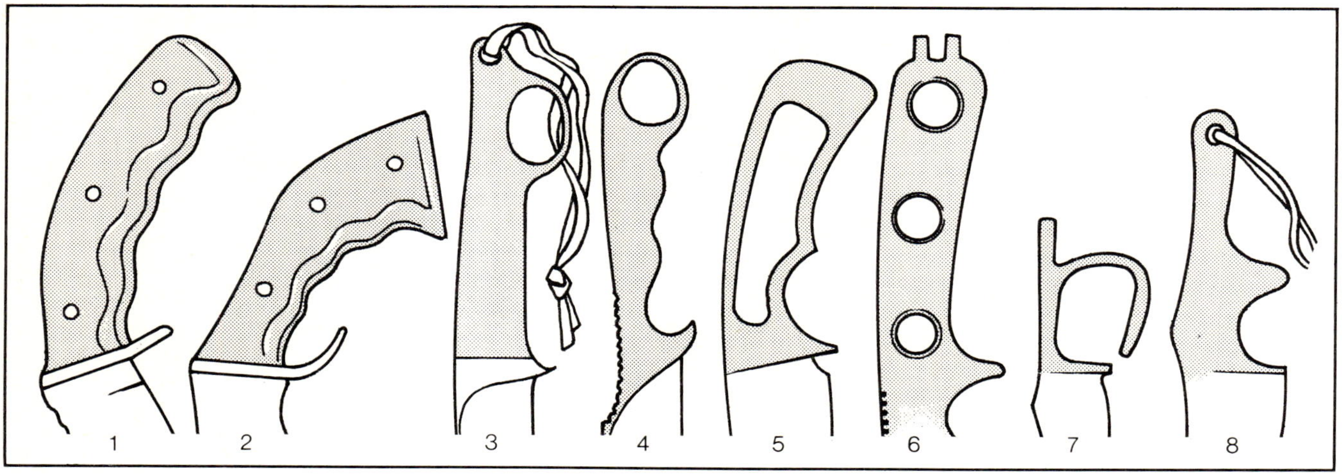

Some peculiar types of handle: 1. curved with finger grips 2. curved pistol grip-all steel 3. handle with protruding bow 4. with end bow 5. hollow 6. a diver's handle 7. hook shaped 8. with single finger grip.

rapier In the Middle Ages, a Franco-Spanish sword used more for dress, with a wide blade, an edge and a sharp point. In the 16th and 17th centuries (Germany and England) a sword with a long, narrow and rigid blade; it was used along with the 'left hander'* or dagger*.

razor See p. 123

revolver-knife See p. 123

reheating A heat treatment for metals which consists in an even reheating followed by a slow cooling. The purpose of the reheating is to augment the resilience of the steel, i.e. its resistance to shocks.

retractable knives Certain penknives, scrapers, etc., whose blades are spring-mounted and which retract into the handle.

ricasso 1. A flat area on the blade in front of the guard which often bears the manufacturer's mark.

ritual razor See p. 156

rivet A part of the assembly of the blade and handle, vaguely resembling a nail but with a head at either end, made of soft steel or a non-ferrous metal.

roast knife See p. 73

round knife See 'dull knife'*

sabre Originally, the sabre and a sword which can both cut and thrust were confused with each other. With time the sword, or épée, was used only for thrusting, and the sabre for both. The form and size of a sabre have varied over the centuries: a straight or curved blade, flat or hollowed, a handle with one or more branches, etc. See p. 121

sacred knife or *sacrificial knife* See p. 123

sailor's knife See p. 118; also fisherman's knife* and yachtsman's knife*

sardine fisherman's knife See p. 121

saddler's knives Saddlers use several different kinds of knives including the foot knife (see knife*); the straight knife, the curved knife, with a point which curves in, and a back which forms two separate curves.

sausage knife See p. 117

saw-knife See p. 72

scabbard A sheath or case to hold the blade of a sword, dagger etc. See p. 34

scales 1. Metal plates which hold the springs and blades of a folding knife, to which they are joined, by pins which also mount the spring and blade. 2. Also used as synonym for covers or grips. See covers*

scaling knife See p. 117

scalloped edge An edge made up of points separated by beveled semicircular indentations.

scalpel The scalpel and the bistoury can hardly be distinguished except by their use — dissection, autopsies, or operations as the case may be. Also see p. 118

scalpel-razor See p. 118

scalping knife A knife used for removing the scalp used by the eastern Indians of North America, the Gran Chaco, and some ancient Siberian peoples. Its shape varied from tribe to tribe but was so made as to cut without without tearing. It now appears to be certain that it was the Europeans who invented scalping, after having noticed that the Indians (like their 'cousins' the Samurai), attached a special importance to their hair. The Indians, eager to learn the whites' ways and their most fearsome tricks, took to scalping in their turn which led them to acquire scalping knives in large quantities.

scarificater A surgical instrument

made up of several spring-loaded lancets.

scout knife See p. 44

scramasax See p. 16

secret knife See p. 96

semispatium See p. 16

serrated edge An edge with teeth like a saw edge, sometimes with sections which alternately face forward and back so that the blade cuts in both directions.

serving knife See p. 71

Sheffield knife See p. 136

shafted weapons See p. 31

sheath A case for the blade of a knife, sword etc. Usually out of leather but can be metal, etc.

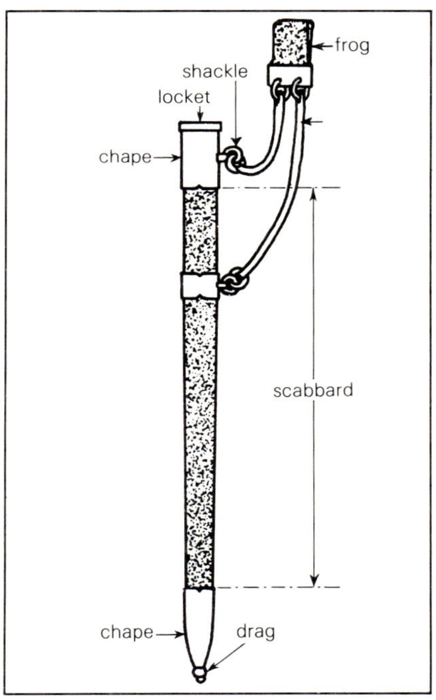

shortsword In the narrow sense, a short Roman sword of Iberian origin. It is a thrusting weapon with two edges.

sica See p. 31

side arms See p. 31

single handed knife See p. 131

skinning knife See p. 44

skiver (skiving knife) See p. 120 and also tanner's knife* and paring knife*

slice 1. A knife with a broad, thin, flexible blade used for cutting and serving food.

slicing knife 1. A medieval knife with a very broad blade. The point was sometimes squared off. If it was pointed it could be used to spear the meat 2. A tanner's knife 3. A kitchen knife.

smoker's knife See p. 126

snow knife See p. 157

spatula Instruments in the shape of a small trowel or dull edged knives, used by various professions: surgeon, mason, printer, painter, druggist, etc.

spear-sword See p. 30

sports knife See p. 41 and following pages

spinner's knife See p. 121

spring A band of steel so made as to return to its original form when not compressed. For knives, three kinds of spring are used: straight springs for ordinary knives, snap springs for knives which snap shut and must be opened by moving a catch, springs with a tenon which is released by pulling on a ring (certain switch-blades).

spring knife See p. 96

spu-gri See ritual razor

steak knife See p. 72

steel, (types of) See pp. 25 and 43

sticker See p. 117

stilleto 1. A very thin dagger with a triangular handle 2. Every knife with a thin blade which recalls this shape.

straight knife See p. 118

striking weapons See p. 30

surgical knives and instruments In the narrow sense, surgical knife is synonymous with the straight knife used in amputations (see p. 118). But the term can be applied to all surgeon's edged instruments and particularly the cataract knife, used for corneal transplants; the Cheseldon or lithotomy knife used for removing gallstones; the interostic knife with two cutting edges used for separating flesh in the interbone areas of the leg or forearm, and the lenticular knife used to reduce irregularities in the cranial bone produced by a trepanation. See also scalpel*

survival knives See p. 42

swedge A bevelled forward part of the back of some blades, but which is not sharp.

Swiss knife See p. 103

sword A long straight steel hand weapon with a point. It always has a guard, but otherwise many variants are possible according to the length, one or more cutting edges, the shape of the shaft, etc.

table knives See p. 72 and following pages

taillebois A large medieval knife used for making 'trenchers' when these were still made of wood. See trencher*

taillepain Medieval knife with a sharp point and smaller size than the taillebois also used to make 'trenchers' when these were in the form of large flat baked bread. See trencher*

tang See p. 21 and illustration p. 168

tanner's knife Tanners use various kinds of cutting instruments such as the skiver to even out and prepare skins.

tanto See p. 155

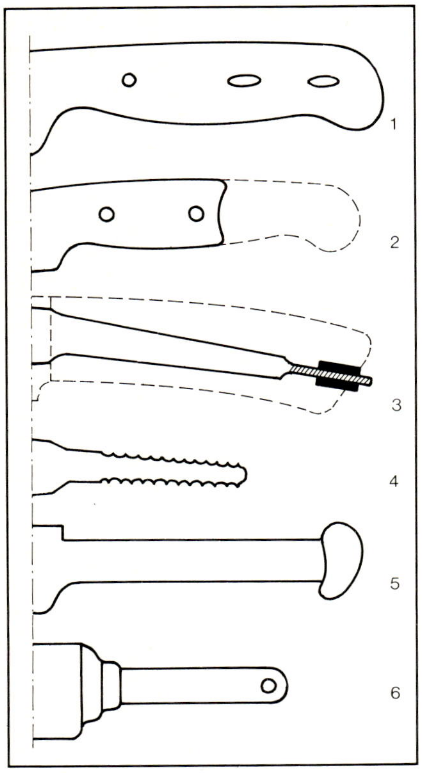

The types of tangs are relatively few. Here are the principal ones (also see page 22): 1. full tang 2. half tang 3. narrow tang with tightening nut 4. ribbed narrow tang 5. partial tang 6. push tang with rivets.

tart-slicer A pastry knife in the shape of a shield with one or two edges, sometimes with fine teeth on one side.

throwing knife The throwing knife can come in several different shapes. In the West where it is used exclusively for sport, only the point is sharp although both sides are bevelled. The blade is quite thick to balance the weight of the handle.

thumb piece A flat part on the spines of some knives, or handles, or again, sometimes even a special folding piece, to rest the thumb upon and give added force and accuracy to the stroke.

tiger knife See kuttar*

toiletry knives See p. 123

tomato knives See p. 72

tomato peeler See p. 72

tranchelard See p. 72

trapper's knife See p. 52

trencher A plate on which meat was cut or served in the Middle Ages; it could be of wood, metal, or a large flat piece of bread serving as a plate.

tripe knife There are several such knives. The most common curves out, the blade narrows towards the point, the end is blunted.

trowel See slicer*

tungsten A heavy metal skin akin to molybdenum. In alloys, it is above all used for fast cutting tools because of the hardness of tungsten carbide.

unusual knives See p. 123 and following pages

utility knives See fighting utility knives*

vanadium A silver-gray metal, the hardest of all. Ferro-vanadium, obtained in the electric furnace by the action of carbon on an anhydride of vanadium and iron oxide makes a refined steel and is also used to prepare special steels.

vegetable knives See p. 72 and following pages

veining knives See p. 117

velvet knife See p. 121

wedding knife See p. 16

weighing-knife See p. 112

well The lower part of a blast furnace (whence its name from its forerunner) where the melted metal collects. The material from which it is made is often clay mixed with cement, but this can vary according to the metal, etc.

whetstone An abrasive stone for sharpening knives or other edged tools.

white weapons See p. 31

wire-cutter See p. 118

woman's knife See p. 157

yachtman's knife See p. 125

yatagan A long Turkish knife — or short sabre — with a very sharp blade making a curve. A combat and execution weapon.

BIBLIOGRAPHY

The list which follows is not exhaustive. Only the most important works have been cited.

American Blade (period.) *passim.*

L'Art de la coutellerie au pays de Langres et de Nogent. Langres 1966.

Richard W. BARNEY and Robert W. LOVELESS, *How to Make Knives.* North Hollywood 1977.

John D. BATES and James H. SCHIPPERS, *The Custom Knife.* 2 vol. Memphis (Tennessee) 1973-1974.

Claude BLAIR, *The James A. de Rothschild Collection... — Arms, Armour and Base-Metalwork.* London/Fribourg 1974.

Clément BOSSON, René GÉROUDET and Eugen HEER, *Armes anciennes des collections suisses.* Lausanne 1972.

Clément BOSSON et al., *Armes de chasse.* Geneva 1969.

Janine BOUET-CAVIN, *Les Couteliers langrois au XVIIIe siècle.* Dijon 1966.

Bulletin de la société historique... de Langres. XIV/1. 1967.

Arthur DAGUIN, *Nogent et la coutellerie dans la Haute-Marne.* Nogent 1878.

A. de DANIEL, *Description complète des fabriques d'épées et de couteaux... à Solingen.* 1802.

Pierre-Louis DUCHARTRE, *Histoire des armes de chasse et de leurs emplois...* Paris 1955.

Roy EHRHARDT, *Encyclopedia of Pocket Knives.* 3 vol. 1973.

Auguste FOUGEROUX de BONDAROY, *L'Art du coutelier en ouvrages communs.* Paris 1872.

Alfred FRANKLIN, *Les Corporations ouvrières de Paris du XIIe au XVIIIe siècle.* not dated.

Alfred FRANKLIN, *Dictionnaire historique des arts, métiers et professions exercés dans Paris depuis le treizième siècle.* Paris/Leipzig 1906.

Ch. GENART, *Coutelier de la fabrique collective de Gembloux* (Belgium). 1892.

Alain-Charles GRUBER, *Kostbares Essbesteck des 16. bis 18. Jahrhunderts.* Berne 1976.

Guns and Ammo Guidebook to Knives and Wedged Weapons. Los Angeles 1974.

Henri HARVARD, *Dictionnaire de l'ameublement et de la décoration depuis le XIIIe siècle jusqu'à nos jours 4 vol.* Paris. not dated.

HENRY, *Précis descriptif sur les instruments de chirurgie anciens et modernes.* Paris 1825.

B. R. HUGHES, *American Handmade Knives of Today.* Union City 1972.

B. R. HUGHES and Jack LEWIS, *The Gun Digest Book of Knives.* Northfield 1973.

Knife World (period.) *passim.*

Abbé LALANNE, *Histoire de Châtellerault et du Châtelleraudais.* not dated.

Roger LALLEMAND (coll.), *Guide du découpage en cuisine.* Cousances 1976.

E.-O. LAMI, *Dictionnaire de l'industrie et des arts industriels.* not dated.

H. C. LANDRIN, *Manuel de coutelier.* Paris. not dated.

H. C. LANDRIN, *Nouveau Manuel complet du fabricant d'instruments de chirurgie.* Paris 1860.

Sid LATHAM, *Knives and Knifemakers.* New York/London 1974.

Maurice LELOIR, *Dictionnaire du costume.* Paris 1951.

Nogent. Bulletin officiel municipal. No 5, 1976.

Camille PAGÉ, *La Coutellerie depuis l'origine jusqu'à nos jours.* 6 vol. Châtellerault 1896-1904. 1616 + CXIV 335 pages of Appendices. Contains lists of cutlers, extracts of corporation statuts, statistics, economic and social studies, a glossary, a lexicon of cutlery terms in 180 languages and dialects, etc.

Jean-Jacques PERRET, *L'Art du coutelier.* 2 vol. Paris 1771-1773.

Harold PETERSEN, *A History of Knives.* New York 1966.

Heinz G. PFAENDER, *Das Tischmesser, eine kulturgeschichtlich-technologische Untersuchung.* Stuttgart 1957.

Manuel RICO Y SINOBAS, *La Cuchilleria en España.* not dated.

Oscar RIS-PAQUOT, *Dictionnaire encyclopédique des marques et monogrammes...* 2 vol. Paris 1893.

Dominique-Gustave SAINT-JOANNY, *Simples Notes pour servir à l'histoire de la ville de Thiers aux trois derniers siècles. I. La Coutellerie thiernoise de 1500 à 1800.* Clermont-Ferrand 1863.

Hugo SCHNEIDER, *Schweizer Griffwaffen.* Bern 1968.

Norman STRUNG, *An Encyclopedia of Knives.* Philadelphia/New York 1976.

Mel TAPPAN (cir.), *A Guide to Handmade Knives.* Rogue River (Oregon) 1977.

H. R. UHLEMANN, *Alte Tischgesellen, vom Fürschneyden, Trincieren und Kredenzen.* Bamberg 1966.

Alfred de VAULABELLE, *Les Couteaux, leur histoire, leur fabrication.* not dated.

VIOLLET-LE-DUC, *Dictionnaire raisonné du mobilier français.* Paris 1855.

Ken WARNER, *The Practical Book of Knives.* South Hackensack (New Jersey) 1976.

John WEISS, *A Catalogue of Chirurgical Instruments Invented and Improvised.* London 1825.

*

* *

Current and back catalogues: Auerhahn, Berndorf, Bone Knife, BSF, Case, Centofante, Chabanne, Christofle, Cooper, Cousances, D'holder, Dowell, Ern, FOB, France-Exportation, Frank, GAB, Gerber (Germany), Gerber (U.S.A.), Granica, Giesser, Gutmann, Haller Geschenk-Import, Hawkins, Herder, Hueske, Jacobs, Jensen, Jezler, Kron, Lee, Little, Loveless, McBurnette, Martor-Argentax, Mellor, Moran, Nolen, Petersen, Pfeilring, Platama, Puma-Werk, Randall, Samson, Sigman, F. L. Smith, John T. Smith, Norsk Stalpress, Thérias, Tiffany, Towell, Victorinox, Wingen, Zack, Zwillingswerk.

Note: When ordering a catalogue from a knifemaker, it is recommended that a check for a few dollars be added to cover costs.

A WORD OF THANKS

Many people have helped to make this book possible by lending pieces out of their collections, providing photographs and documents, information and advice. For those who cannot be named, the author and publisher here express their cordial thanks.

Particular thanks are extended to the following:

Mr. Robert Barincou, collector, Paris; MM. Bauermann and Cegielski, of the J.A. Henckels Zwillingswerk AG, Solingen (Germany); Mr. Gilles Béguin, Keeper of the Musée Guimet, Paris; Mr. Ernest Berger, collector, Lausanne; Mr. Gabriel Besson, cutler, Lausanne; Mr. Kemal Cig, Keeper of the Topkapi Sarayi Müzesi, Istanbul; Mr. Paul-Louis Frachebourg and Mrs. Monique Fillettaz, of the Coutellerie du Mont-Blanc, Geneva; Mr. H.H. Frank, knifemaker, Whitefish (Montana); Mr. Marwan Hamadé, "An-Nahar", Beyrouth; Mr. Karl-Heinz Hauptreif, Director of the Klingenmuseum, Solingen (Germany); Mr. André Jeanneret, Director of the Musée d'ethnographie, Geneva; Dr. Hermann Jedding, of the Museum für Kunst und Gewerbe, Hamburg; Dr. Roland Klemig, Keeper of the Bildarchiv Preussischer Kulturbesitz, Berlin; M. le Marquis de Lastic, Keeper of the Musée de la Chasse, Paris; Mr. Harvey McBurnette, knifemaker, Piedmont (Alabama); Swiss National Museum, Zurich; The House of Eloi Pernet, Nogent and Paris; Mrs. Margot Raissac, Keeper of the Musée Bouilhet-Christofle, Saint-Denis; Mrs. Michèle Schillé, Head of the Service commercial des Fonderies de Cousances, Ancerville; Miss Hanni Schwab, County archeologist, Fribourg (Switzerland); Mrs. Germaine Tureau, of the Réunion des musées nationaux, Paris; Mrs. Geneviève Veyssière, Head of the service d'information of the Musée des arts et traditions populaires, Paris; Colonel A.R. Waller, Administrator of the National Trust for Places of Historic Interest, Aylesbury (Great Britain); Mr. Buster Warenski, President of the US Knifemakers' Guild, Richfield (Utah); The House of Anton Wingen Jr., Solingen (Germany); Mr. Egon Winter, Puma-Werk, Solingen (Germany); Mr. Don Zaccagnino ("Zack"), knifemaker, Pahokee (Florida).

We also wish to thank:

Mr. Charles Bernard, cutler, Lausanne; Mr. Clément Bosson, Keeper of the Musée d'art et d'histoire, Geneva; Mr. J.F. Brugière, of France-Exportation, Thiers; Mr. A.K. Corry and Dr. A.K. Newmark, Assistant keepers of the Science Museum, London; Mrs. Mary F. Doherty, of the Metropolitan Museum of Art, New York; Ecole polytechnique fédérale of Lausanne; Mr. C. Elsener, of the Victorinox Messerfabrik Elsener, Ibach (Switzerland); Mr. Rade Hawkins, knifemaker, Red Oak (Georgia); Mrs. N. Hervoir, Municipal librarian, Châtellerault; Mr. A.I. Hofmann, of the "Deutsches Waffen-Journal", Schwäbisch-Hall (Germany); Mr. Milo ("Mike") Leach, knifemaker, Swartz Creek (Michigan); Mrs. Ann-Marie Lönnberg, of GAB, Stockholm; Mr. Dominique Minel, of Victoria Nogent S.A., Biesles; Mr. William Moran, knifemaker, Braddock Heights (Maryland); The House of Norsk Stalpress AS, Bergen (Norway); Dr. Annaliese Ohm, Keeper of the Museum für Kunsthandwerk, Frankfurt-am-Main; Dr. Ch. Osterwalder, of the Bernisches historisches Museum, Bern; Mr. Maurizio Palaci, of Bruno Danese SNC, Milan; Mr. J.F. Russell, Research assistant at the British Museum, London; Mr. Vésale van Ruymbeke, collector, Lausanne; Mr. Corbet R. Sigman, knifemaker, Liberty (West Virginia); Mr. Gérard Tisserand, Keeper of the Musées des Langres; Mrs. Fernanda Matilda Revuelta Tubino, Director of the Museo de Santa Cruz, Toledo (Spain).

PHOTOGRAPHIC CREDITS

Only particular cases are included here. For the sources of photographs or documents see the captions of the illustrations in question.

t — top	m — middle	b — bottom
l — left	c — center	r — right

Antiquités nationales, Saint-Germain-en-Laye/Giraudon: 6. — Musée du Louvre, Paris/Giraudon: 8. — Museum für Kunst und Gewerbe, Hamburg/Karin Kiemer: 11 l and c, 12 b, 18 r, 81 l, 89 br, 91 r. — National Trust for Places of Historic Interest, Aylesbury (Great Britain): 13 b, 16, 19 tl, 31, 32, 35 r. — Bayerische Verwaltung der Staatlichen Schlösser, Gärten und Seen, Munich: 14, 84 tl, 86 bl, 120 r, 126 t, 130 bl, 148, 150 tc. — City Museum, Sheffield/West Park Studios: 15 t, 17, 20, 25, 27 r, 36 r, 70, 80, 83 tl, 84, 85 b, 86 br, 88 t, 94, 96 r, 97, 98, 99, 100, 101 t, 102, l04, 120 l, 122 t, 126 b, 127, 128, 129, 130, 132, 133, 134 tl, 135 r, 136, 139 tl, 142 t, 151 tl, 154, 155 r. — Edita, Lausanne: 21, 22 tr and b, 23 tl, 26 m, 27 bl, 30, 34, 40, 72-73 t, 75, 96 l, 107 b, 114 b, 118, 119, 159 sq. — Coutellerie du Mont-Blanc, Geneva/Edita: front cover, 37, 38, 39, 51 tr, 54 tr, 101 br, 103 t, 113, 114, 115, 122 b, 125, 131, 134 b, 135 bl, 138, 139 tr and b, 142 b. — Yvan A. de Riaz, Nyon (Switzerland)/Edita: 111 b. — Jean-Rodolphe Piccard, Treytorrens (Switzerland) and Coutellerie du Mont-Blanc/Edita: 140.— Zwilling-Werk, Solingen: 22 tl and tc, 23 (except tl), 27 tl, 73 b, 76 l, 77, 78, 121 t. — Randall Knives, Orlando (Florida): 56, back cover. — Réunion des Musées nationaux, Paris: 123, 124, 141, 147, 150 b, 151 tr and b, 152 t, 153. — Musée d'ethnographie, Geneva: 143 r, 145, 150 tl and tr, 152 b, 155 l, 156, 157, 158; (gift, J. Détry): 143 l; (gift, J. Bouvard): 143 c; (gift, A. Hauenstein): 144.

This book was published under the direction of
Ami Guichard
Produced under the direction of Charles Riesen
Designed by Max Thommen
Printed by G.E.A., Milan
Bound by Maurice Busenhart, Lausanne

Printed in Italy